Future-oriented studies

The three objectives of the transverse programme *Future-oriented studies* are as follows:

to highlight the future developments in UNESCO's fields of competence;

to strengthen UNESCO's clearing-house function in the field of future-oriented studies;

to promote training activities in future-oriented research and studies.

Trois objectifs ont été assignés au programme transversal *Études prospectives :*

mettre en relief les faits porteurs d'avenir dans les domaines de compétence de l'UNESCO ;

renforcer la fonction de centre d'échange d'information de l'UNESCO dans le domaine des études prospectives ;

promouvoir des activités de formation en matière de recherche et d'études prospectives.

Biotechnologies in perspective:

socio-economic implications for developing countries

Edited by Albert Sasson
and Vivien Costarini

UNESCO

The designations employed and the presentation of the material in this publication do not imply the expression of any opinion whatsoever on the part of the UNESCO Secretariat concerning the legal status of any country or territory, or of its authorities, or concerning the delimitations of the frontiers of any country or territory.

Published in 1991 by the United Nations
Educational, Scientific and Cultural Organization
7 Place de Fontenoy, 75700 Paris
Composed by UNESCO
Printed by Presses Universitaires de France, Vendôme

ISBN 92-3-102738-7

Foreword

In the opinion of many specialists, biotechnologies have in general kept their promise with regard both to results expected from research and to their potential applications over the last ten years, in particular in the industrialized countries. The last decade of the twentieth century will without doubt witness the marketing of an increasing number of biotechnology-derived products, which will co-exist with conventional products or will replace them. Reflection on the economic and socio-cultural implications of the commercialization of these new products deserves to be developed at regional and international levels by the greatest number of people and institutions concerned, in particular from developing countries.

During recent years, emphasis has been placed on ethical issues concerning biotechnologies, which have been the subject of reflection by national ethics committees, governmental or intergovernmental public institutions, and non-governmental organizations, with a view to paving the way for legal texts.

It is, of course, true that genetic engineering is used in biotechnologies, as well as in embryology and medicine, and that the release of genetically engineered micro-organisms or plants raises ethical issues and lends itself to debate on the effects of this release on the environment. Nevertheless, this bio-ethical reflection does not directly concern biotechnologies strictly speaking, i.e. the utilization of the synthesis and transformation ability of living cells – microbial, animal and plant – cultivated on particular media and producing a wide range of useful substances.

The International Seminar on the Economic and Socio-cultural Implications of Biotechnologies, held at Vézelay, France, from 28 to 31 October, 1990, did not come within this framework of bio-ethical reflection, but concerned the social, economic and cultural implications of biotechnological innovations, adopting a future-oriented approach, in which past and present situations throw light on the future.

Several types of positive or negative impact of biotechnologies on developing countries are conceivable:

- on the improvement of crop productivity and overall food production through the direct use of plant biotechnologies, as a complement to conventional breeding techniques;
- on nutrition, through improvement in the quality of agricultural and agro-industrial production, and the up-grading of food fermentations;
- on the better integration of food production with bio-energy production and consumption at household and village levels;
- on the increase and improvement of livestock production and the health of domestic animals;
- on the accurate diagnosis and prevention of diseases, and on the improvement of public health;
- on trade patterns between developing countries and industrialized countries resulting from differences in the timing of productivity increases both in agriculture and in agro-industry between the two groups of countries, and from the commercialization of new bio-industrial products that tend to replace foodstuffs and various biochemicals produced by the developing countries, thus depriving the latter of an important source of income (displacement effects);
- on income and employment;
- on the possible extension of cash crops at the expense of food crops;
- on the strengthening of the large agricultural estates, resulting in the displacement of small landholders and landless farmers;
- on the possible decrease of genetic diversity as a consequence of the widespread distribution of new cultivars;
- on the increased privatization of research results in plant biotechnologies which entails, for developing countries, more difficult access to these results and the payment of fees for the seed of crop varieties bred in industrialized countries.

The purpose of the seminar was to bring together specialists, mainly from the social and human sciences, to analyze the economic and socio-cultural implications of biotechnologies as applied to agriculture, livestock husbandry, food and energy production, and human and animal health. The participants took stock of available knowledge, identified gaps in research and suggested directions for further investigation. It is hoped that the dissemination of the conclusions of the seminar and the publication of a number of position papers will contribute to a better understanding of the impact of biotechnologies in these fields, and will encourage the competent institutions to take account of these considerations in drawing up economic and social development policies, so as to increase the positive impact of the expansion of biotechnologies and decrease the negative effects.

The seminar was organized by the Division of Studies and Programming, which is responsible for the implementation of UNESCO's transverse programme, "Future–oriented studies", in collaboration with the Organization's Science and Social and Human Sciences Sectors. Inasmuch as reflection on the economic and socio–cultural implications of biotechnologies is of great interest to developing countries, but has, for the most part, been undertaken in industrialized countries or by international bodies, the participants – twenty in number – were mostly from developing countries.

Contents

Acknowledgements

UNESCO expresses its deep gratitude to Mrs Calliope Beaud, the co-organizer of the seminar, who contributed largely to the success of the meeting, held in the town of Vézelay, which provides a historic setting for the meetings of international repute organized by Mrs Beaud on themes such as environment, development and the rights of humankind.

UNESCO is also deeply grateful to all those who contributed to the seminar, whether as authors of position papers or as participants, of whom the list is given below.

Dr I. Ahmed, Technology and Employment Branch, Employment and Development Department, International Labour Organisation (ILO), 4, route des Morillons, CH–12 Geneva 22 (Switzerland)

Dr M. Ben Said, Ministère de l'Agriculture, Institut national agronomique de Tunisie, 43, avenue Charles Nicolle, 1002 Tunis – Belvédère (Tunisia)

Professor P.M. Bhargava, Centre for Cellular and Molecular Biology, Hyderabad 500 007 (India)

Professor S. Bhumiratana, Department of Chemical Engineering, Faculty of Engineering, King Mongkut's Institute of Technology, Thonburi, Bangkok 10140 (Thailand)

Mrs C. Brenner, Development Centre of the Organisation for Economic Co-operation and Development (OECD), 94, rue Chardon-Lagache, 75016 Paris (France)

Dr E.J. Da Silva, Division of Basic Sciences, UNESCO, 7 place de Fontenoy, 75700 PARIS (France)

Dr D. Deo, Department of Sociology, Social Work and Corrections, Morehead State University, Morehead, Kentucky 40351 (USA)

Professor P.F. Greenfield, Department of Chemical Engineering, University of Queensland, St. Lucia, Queensland 4067 (Australia)

Professor Hong–Ik Chung, Graduate School of Public Administration, Seoul National University, San 56–1, Shimlim–dong, kwanag–ku, Seoul (Korea 151–742)

Dr G. Junne, Universiteit van Amsterdam, Faculteit der Politieke en Sociaal–Culturele Wetenschappen, Oudezijds Achterburgwal 237, 1012 DL Amsterdam (Netherlands)

Professor M. Kenney, Department of Applied Behavioral Sciences, University of California, Davis, California 95616 (USA)

Dr M. Leopold, Département de Sociologie, Université du Québec, Case postale 8888, succursale A, Montréal, Québec H3C 3P8 (Canada)

Dr A. Martel Penate, AGROPLAN, Avenida Casanova, Edificio Inca, Apartamento 52, Caracas (Venezuela)

Mr J. Mugabe, African Centre for Technology Studies, P.O.Box 45917, St. George's House, 4th floor, Parliament Road, Nairobi (Kenya), and ACTS Biopolicy Institute, Project Development Office, IFIAS–Maastricht, Witmakersstraat 10, 6211 JB Maastricht, Netherlands

Dr G. Otero, Department of Spanish and Latin American Studies, 8310 CC – SLAS, Simon Fraser University, Burnaby, B.C. (Canada V5A 1S6)

Dr L. Rios, Centro Tecnológico de Antioquia, Edificio EDA 2°, piso planta baja, Oficinas Naciones Unidas, Medellin (Colombia)

Dr I. Robertson, Department of Crop Science, Faculty of Agriculture, University of Zimbabwe, P.O.Box MP 167, Mount Pleasant, Harare (Zimbabwe)

Professor S.L.M. Salles Filho, UNICAMP, Universidad Estadual de Campinas, Caixa Postal 1170, 13100 Campinas SP (Brazil)

Dr A. Sasson, Director, Bureau of Studies, Programming and Evaluation, UNESCO, 7, place de Fontenoy, 75700 Paris (France)

Mr F.C. Sercovich, Regional and Country and Studies Branch, Department for Programme and Project Development, D–2010, United Nations Industrial Development Organization (UNIDO) Vienna International Centre, P.O.Box 300, A–1400 Vienna (Austria)

Mme C. Vincent, *Le Monde,* 15, rue Falguière, 75501 Paris Cédex 15 (France)

Dr P. Weiss, Bureau of Studies, Programming and Evaluation, UNESCO, 7, place de Fontenoy, 75700 Paris (France)

Dr Xu Zhao-xiang, National Research Centre for Science and Technology for Development, P.O.Box 3814, Beijing 100038, (People's Republic of China)

Position papers

Implications of the use of biotechnologies in developing countries, with special reference to India

P.M. Bhargava and C. Chakrabarti

Implications of the use of biotechnologies

POPULATION CONTROL, FOOD AND ENERGY

It will be at least two decades before there is any likelihood of stabilizing the total world population; during this period most of the population growth will occur in the developing countries in Asia, Africa and Latin America; it has been demonstrated that increase in population is related to social, economic and educational deprivation which is greatest in these regions. Biotechnologies would play an important role in reducing this deprivation, and could also be increasingly useful in meeting the growing food requirements and improving the nutritional status of the world population. New techniques such as radiation–sterilization or genetic engineering would increase the shelf–life of food at room temperatures. A decrease in the cost of essential nutrients such as amino–acids and vitamins produced through fermentation procedures using genetically engineered organisms and better recovery techniques, would make it possible for optimal nutritional requirements to be met through pills which would be cheaper than conventional foodstuffs. Plantations using fast–growing trees, well adapted to climate and soil conditions, would provide good firewood and hopefully lead to other productive activities.

AGRICULTURE AND INDUSTRY

If the current trend of migration to the large cities is to be halted and the basic structure of villages maintained in a country like India, it will be important to have the income derived from agriculture supplemented by income from other sources. Biotechnologies could increase industrial productivity while maintaining involvement in agriculture or agricultural-related activities. A judicious use of biotechnologies could lead to a situation in which a large proportion of the population of a country that is primarily agricultural today, continues to be engaged in agriculture but derives a large portion of its income from non-agricultural activities, unlike in the industrialized countries where only a small percentage of the population is engaged in agriculture and produces food for the whole country. Biotechnologies could thus be utilized to bridge the gap between industry and agriculture, and this interaction is bound to give a boost to agro-based industry.

Agricultural practice would also undergo a major change when it becomes possible to have non-leguminous plants fix their own nitrogen and thus obviate the need for fertilizers that are produced today through energy-intensive processes. Similarly, when pest-resistant plants are developed through genetic engineering techniques, the problems relating to pesticide and herbicide residues which, *inter alia,* lead to an increase in the cost of production and processing of food, would be substantially mitigated, thus releasing capital for other uses.

Conventional factories today are large establishments operating on the basis of economy of scale. Biotechnology-based factories, on the other hand, can be small establishments suitable for villages. Several biotechnologies are labour- intensive and economy of scale is not as important as in chemistry-based industries. Biotechnologies are, as a rule, less pollutive than chemical technologies and the cost of maintenance of biofactories would therefore be much lower. This could well lead to a reduced rate of interest and a lower level of inflation when biotechnologies replace chemical technologies to a substantial extent.

NEW FEARS

Biotechnologies provide a means of domination that may not be immediately obvious, and great vigilance is required in order to avoid misuse of such powerful techniques.

A further problem would be that of contamination of products manufactured through biotechnologies. For example, if human insulin can be made cheaply and efficiently in *Escherichia coli,* steps would have to be taken to ensure that the insulin administered to human beings does not carry with it any *E. coli* protein, which could act as an immunogen and cause problems. Since the immune response can be elicited by extremely small amounts of a material, new methods of purification of genetically engineered products and new techniques of quality control would have to be evolved.

There would also be the problem of checking and regulating the release of new organisms, variants and species – microbial, plant or animal – obtained through biotechnologies such as somatic cell hybridization or genetic engineering. The ability would have to be acquired to make a reasonable prediction as to how such release would alter the environment.

Advantages for developing countries

Biotechnologies could be major tools for providing additional employment and sources of income for those engaged in agriculture, especially in countries where agriculture has been so far the main occupation of the majority of people: thus they would help to effect a gradual transformation from an agricultural economy to an industrial economy without the concomitant disadvantages of urbanization or migration from villages to large cities.

Other advantages would include the provision of cheap, reliable sources of firewood through plantations, thus solving at least a part of the energy problem in developing countries, the increase in food production, better health (for example, by making important diagnostics, medicines and vaccines available at a much cheaper rate than is possible today), and a cleaner environment.

Assets of developing countries and optimal use

Many developing countries have certain advantages regarding biotechnologies, in comparison with the developed countries. India, for example, is well placed today as its level of scientific advancement is now one of the highest among the developing countries. It produces a wide range of consumer goods and exports, highly sophisticated finished products made entirely in India; its engineering and construction firms have won a large number of major contracts abroad in competition with multinationals; it has an elaborate infrastructure for science and high-level expertise in a whole range of areas in science and technology; and it is the only developing country which publishes more than 10,000 papers a year in scientific journals. Although the quality is not always high, India has a large number of trained biologists in institutions spread throughout the country.

In the past five years, the Government of India has given top priority to basic research in biology and development of biotechnologies. India has a strong organized public sector as well as a private sector, both of which have a tradition of scientific entrepreneurship. It has active professional societies in the area of biology, and its biologists are aware of developments elsewhere, which means that conclusions may be drawn from the successes and failures of other countries. The country is rich in natural resources. Many of the above advantages or assets would apply to other developing countries.

In order to ensure that biotechnologies do not become a tool of exploitation but are used universally for the benefit and progress of humankind, it may be necessary to:

- increase awareness concerning modern biology and biotechnologies, especially the current and likely advances, on a world scale;
- ensure that there is no obstacle to information relating to biotechnologies, and that both developed and developing countries follow a global policy in this respect;
- promote the free exchange of information between countries in respect of biotechnologies, not only among specialists but also at government level;
  implement appropriate confidence-building measures on a world scale.

References

BHARGAVA, P.M. 1989. The Biological Revolution: What we need to do about it. Silver Jubilee Lectures, Collected volume, India, Mysore, S.J. College of Engineering.

BHARGAVA, P.M. 1989. Social Implications of Modern Biology. *India International Quarterly,* Spring 1989.

BHARGAVA, P.M. 1990. *Conventions in Place of Patent Laws.* Mainstream. March 1990.

BHARGAVA, P.M.; CHAKRABARTI, C. 1990. The Role and Present Status of Biotechnology in India. *Current Science* (in press).

CHAKRABARTI, C. 1989. Biotechnology: the Basis and the Prospects. *Journal of Scientific and Industrial Research,* No. 48.

Implications of biotechnologies for third world agriculture: lessons of the past and prospects

S.D. Deo

Introduction

Modern agriculture, based on science, has produced an abundance of food in some developed countries. It can be argued that this has been facilitated not by science *per se* but by economic forces. It is true that overproduction of food would not have been possible without the technological means to exploit the agricultural land. Yet it is necessary to remember that agricultural technologies are part of dialectical processes constrained by the changes in social relations within political economies (Levins and Lewontin, 1985). Any discussion of biotechnologies in the context of agriculture or any other area must articulate the political and economic objectives of their development and research as a means to historically conditioned social ends.

It is argued that the ascendancy of biotechnologies and genetic engineering in agriculture has come about not merely because they are powerful new techniques of production in the hands of industrialized nations, but because those nations wish them to be seen as having a talismanic capacity. Unless there is a promise of solutions to social, economic, and political problems, there will be no faith in science and technology. This provides an opportunity to examine the context of science and technology, development and modernization, agricultural development and the current precarious situation of Third World countries, in order to search for new and different ways of achieving the advancement of these societies.

Development issues

The popular concept of modernization was presented as a historical process in which Third World countries lagged behind the advanced industrial nations which were also the former colonial powers. The way out of the predicament in which these new nations found themselves was to follow the proven path laid down by the western nations. This emulation of developed countries was promoted by economists and modernization theorists. This development approach enabled normalization of the world and led to new forms of domination (Escobar, 1984–1985). The age of development gave rise to new set of institutions, and policies and programmes based on such notions as 'modernization = industrialization = westernization', the theory of 'comparative advantage', and the unavoidable dependence on science and technology.

The comparative advantage theory posited a world of variegated and unequally distributed resources, the exploitation of which each nation could successfully practise in an international division of labour which would eventually lead to stable economic growth. It is not surprising that underdeveloped nations were expected to continue concentrating on agricultural production and research on export-oriented commodities. The leaders of these countries were more than receptive to the idea that in order to industrialize rapidly, a large surplus, both real (food and raw materials) and financial (in the form of savings) must be generated. In order to do so they had to depend on new international markets and to maintain and protect their traditional market share, thus obtaining foreign exchange with which to acquire industrial technology and know-how.

These policies have not improved the situation of Third World countries. The colonial pattern of using the best available land for commercial crops and marginal land for food crops continued. The production of basic foods was significantly less than that required to feed rapidly-growing populations. Aid as an alternative to, or supplement for, internal resources proved an expensive and humiliating experience to most Third World countries.

The material success of the western industrial nations, presumed to be a product of scientific effort, has provided an attractive model for the elites in the Third World to follow. The Third World countries have desperately sought to catch up with the developed nations by importing western science and technology in the hope of improving their living conditions and liberating their societies from neo-colonialism.

The development of agricultural sciences has been traced to the establishment of botanical gardens, agricultural experiment stations, the expansion of colonial empires, and the shift from subsistence farming to capitalist agriculture (Brockway, 1979; Busch, 1984; Levins and Lewontin, 1985). L. Brockway's work (1988) on the activities of Royal Botanical Gardens at Kew and their role in the expansion of colonial empires based on, *inter alia,* the development of several highly profitable and strategically important plant-based industries in tropical colonies is important. The initial connection between agriculture, agricultural research and the expanding world capitalist economy is observed in the work of botanical gardens and experiment stations (Brockway, 1979; Busch and Sachs, 1981). This tradition continues in the work of international agricultural research centres – IARCs (Mooney, 1983). The activities of botanical gardens enabled the colonial powers of Europe to develop a genetic foundation for production of plantation crops and cheap food (Kloppenburg, 1988).

Farmers in Third World countries had developed agricultural practices and technologies over a long period of time. Their profound understanding of conventional practices enabled them to raise crops on a consistent basis. When colonial policies forced plantation and other commercial crops on India, agricultural research concentrated not on domestic food crops, but on exportable crops. The resulting agricultural system could neither provide an adequate grain production nor generate a large enough surplus for industrial development. As a result, the agricultural research system largely replicated colonial institutional structures and was susceptible to bureaucratic controls (Arnon, 1968; Anthony *et al.,* 1979; Bhaneja, 1979). The purpose of agricultural educational institutions in India, as in many other countries, was to train management personnel capable of controlling land and revenue for the state. In the recent past, when Western experience in the form of land-grant institutions was exported to many countries, it tended to reinforce the old colonial pattern by replacing one set of technologies and institutional structures by another. The purpose was to facilitate expansion and control of national and international capital, while precluding the welfare and interests of the multitude of small, marginal farmers, as well as rural and urban working classes (Valianatos, 1976; Dahlberg, 1979).

When 40% of value is added in agricultural inputs (fertilizers, machinery, seeds, hired labour, fuel, pesticides) and 50% is added in processing and transportation of the farm output, the thrust of agricultural research would not be towards the needs of small farmers. Science and technology have been used as substitutes for the natural processes (Goodman *et al.*, 1987). Each time such substitution took place the objective was to increase the market share, appropriate more surplus and/or increase control over natural processes and products by making farmers more dependent on purchased inputs (Byres, 1981; Vandermeer, 1981). The process has been described as industrialization of agriculture. Biotechnologies enable natural substances available in the biomass to be transferred at low energy cost and on a larger scale into a variety of materials for use in all sectors of the economy where organic chemicals are produced and used (Redclift, 1984). The promise of cost reduction, efficient production, reduction in dependence on sources of raw materials from outside, and most importantly, the use of patent laws to maintain proprietary rights, makes biotechnologies attractive. It is doubtful that this promise means much to small farmers, i.e. to an overwhelming proportion of the population in the Third World.

CONSERVATION AND USE OF PLANT GENETIC RESOURCES

The displacement and loss of conventional varieties because of increasing acreage under hybrids and so-called high-yielding varieties during the "green revolution" has resulted in a rapid erosion of genetic diversity (Safeeulla, 1977; Mooney, 1980 and 1983; Reichert, 1982; Wilkes, 1983; Doyle, 1985; Kloppenburg, 1988). This erosion threatens the availability of genetic resources for future breeding of crops. The agricultural research establishment in Third World countries is once again unable to find socially relevant solutions. India can claim to have the third largest scientific and technological manpower pool, but that pool remains a large and fragile community (Shiva and Bandopadhyay, 1980; Goonatilake, 1984). The need for protection of agriculture puts "gene-poor" western countries in a position where mutual trade-offs need to be negotiated with "gene-rich" Third World countries. Moreover, plant genetic resources as free goods from the Third World have been worth untold billions of dollars to the advanced capitalist nations (Mooney, 1980 and 1983; *New Scientist*, 1982; Kloppenburg, 1988). The wild relatives of cultivated varieties as well as the traditional varieties are the raw materials for biotechnologies and genetic engineering; without them, new varieties cannot be created through genetic manipulation. If Third World countries are not careful, this could create problems similar to those seen during and after the "green revolution" but with greater environmental uncertainty.

PATENTING ISSUES

The entry of multinational corporations into agricultural biotechnologies was facilitated by the extension of the general patent laws to asexually reproduced plants and new life forms. The existing markets for seeds and fertilizers are large enough to attract capital investment. It is estimated that the agribusiness market in the USA will increase to $ 47.9 billion by 1995. During 1990–1995, the biotechnology component is expected to increase its share in this market from $ 0.5 billion to $ 2.3 billion (Klausner, 1986). The Plant Varieties Protection Act (1970) enabled the private seed companies to patent their varieties and thus initiated the privatization of seed development and distribution (Kloppenburg, 1988). The takeover of seed companies by petrochemical and pharmaceutical multinationals to channel their proprietary products has led to an alarming concentration of control in agribusiness.

The Plant Breeders' Rights subject Third World countries to a new technological treadmill. New varieties will be released not because they represent clearly better material, but in order to increase prices, to protect expiring patents or to secure higher profits for the multinational corporations. Berlan and Lewontin (1986) have argued that private breeders have a vested interest in reducing the lifetime of varieties as far as possible, so that farmers will adopt new varieties every year.

Some issues which Third World countries should deal with realistically are: the role played by patent protection; pressures exerted on Third World countries to change their patent legislation; non-recognition of research results outside the USA for granting of U.S. patents (Coleman, 1988); and the use of 'prophetic' patents, based on hypotheses, library research and discussions rather than on laboratory work. There is a tendency to transform public research into private property and what was public and open has now become private business (Dickson, 1984).

SUBSTITUTION OR DISPLACEMENT OF COMMODITIES

An important consequence of the use of biotechnologies in agriculture is the possible effect on trade of commodities, especially raw materials. Interchanging different raw materials or increasing the range of agricultural products being used as raw materials tend to lead to a restructuring of world markets. Developing countries have long been identified as raw material producers. Yet, the application of biotechnologies can make industrialized countries less dependent on these imports. This is one of the reasons why extensive resources are poured into biotechnologies. While this may be advantageous for the industrialized world, it is disastrous for most developing countries exporting raw materials. See the chapter by Junne.

It is essential to develop a clear idea of the impact of biotechnologies and to devise ways and means of assessing it. At the same time it is necessary for people of Third World nations to guard against conceiving the issues in narrow nationalist terms. Technological choices are first and foremost political ones, and their effects convey the state of relations among social forces (Touraine, 1988).

Conclusions

There is no doubt that biotechnologies offer a wide array of techniques that are flexible and applicable in many areas, but it is also true that they have not emerged in a vacuum nor are they neutral in their socio-economic and political consequences. The pace of development of these techniques has been fast and the technology-optimistic literature has conditioned our thinking in a way which makes it difficult to ask the right questions about problems, possibilities and profits. Developing countries should not assume that they can pick and choose from the technology supermarkets of western countries. Instead, they need to explore actively the possibilities of co-operation and collaboration among themselves, pooling their natural, scientific, technological and manpower resources to define problems and seek solutions that will benefit the weakest segments of their societies. These countries should simultaneously identify conventional technologies which may not be capital-intensive but can be enhanced to make them relevant to the needs of the twenty-first century. Problems such as long-distance control through privatization of knowledge can be countered if international organizations act as clearing houses for information needed by Third World countries.

References

ANTHONY, K.R.M.; JOHNSON, B.F.; JONES, W.O; UCHENDU, V.C. 1979. *Agricultural Change in Tropical Africa.* Ithaca, N.Y., Cornell University Press.

ARNON, I. 1968. *Organization and Administration of Agricultural Research.* Amsterdam, Elsevier.

BHANEJA, B. 1979. Parliamentary Influence on Science Policy in India. *Minerva,* pp. 70–97.

BERLAN, J.P.; Lewontin, R. 1986. Breeders' Rights and Patenting Life Forms. *Nature,* 322, pp. 785–8.

BROCKWAY, L. 1979. *Science and colonial expansion: the rise of the British Royal Botanical Gardens.* New York, Academic Press.

——. 1988. In: Kloppenburg,J. Jr (ed.). *Plant Science and Colonial Expansion: The Botanical Chess Game in Seeds and Sovereignty; The Use and Control of Plant Genetic Resources.* Durham, N.C., Duke University Press.

BUSCH, L. 1984. Can Agronomy Feed the World? Agricultural Research Policy and World Hunger. In: Ehrensaft, P.; Knelman, F. (eds.). *The right to food,* Montréal. The Canadian Associates of the Ben-Gurion University of the Negev.

——. ; SACHS, C. 1981. The Agricultural Sciences and the Modern World System. In: Busch, L. (ed.). *Science and agricultural development.* Totowa, N.J., Allenheld and Osmun.

BYRES, T.J. 1981. The New Technology, Class Formation and Class Action in the Indian Countryside. *Journal of Peasant Studies,* 8 (3), pp. 405–54.

COLEMAN, H.D. 1988. Protecting non-U.S. Research in the U.S. *Bio / Technology,* 6, pp. 791–2.

DAHLBERG, K.A. 1979. *Beyond the Green Revolution: The Ecology and Politics of Global Agricultural Development.* New York, Plenum Press.

DICKSON, D. 1984. *The New Politics of Science.* New York, Pantheon Books.

DOYLE, J. 1985. *Altered Harvest: Agriculture, Genetics and the Fate of the World's Food Supply.* New York, Viking Press.

ESCOBAR, A. 1984–1985. Discourse and Power in Development: Michel Foucault and the Relevance of his Work to the Third World. *Alternatives,* 10, pp. 377–400.

GOODMAN, D.; SORJ, B.; WILKINSON, J. 1987. *From Farming to Biotechnology: A Theory of Agro-industrial Development.* Oxford, Basil Blackwell.

GOONATILAKE, S. 1984. *Aborted Discovery: Science and Creativity in the Third World.* London, Zed Press.

KLAUSNER, A. 1986. Farmers Can't Succeed without Agribiotech. *Bio / Technology,* 8, p. 759.

KLOPPENBURG, J. JR. 1988. *First the Seed: The Political Economy of Plant Biotechnology 1490–2000.* Cambridge University Press.

LEVINS, R.; LEWONTIN, R. 1985. *The Dialectical Biologist.* Cambridge, Mass., Harvard University Press.

LEWIN, R. 1982. Funds Squeezed for International Agriculture. *Science,* 218 (4575), pp. 866–7.

MOONEY, P.R. 1980. *Seeds of the Earth: A Private or Public Resource. International Coalition for Development Action.* London and Ottawa, Inter Pares.

——. 1983. The Law of the Seed: Another Development and Plant Genetic Resources. *Development Dialogue,* 1–2.

NEW SCIENTIST. 1982. *Foreign Fields Save Western Crops.* No. 96. (October 28), p. 218.

REDCLIFT, M. 1984. *Development and the Environmental Crisis: Red or Green Alternatives?* London, Methuen.

REICHERT, W. 1982. Agriculture's Diminishing Diversity: Increasing Yields and Vulnerability. *Environment,* 24 (9), pp. 6–11, 39–43.

SAFEEULLA, K.M. 1977. Genetic Vulnerability: the Basis of Recent Epidemics in India. *Annals of the New York Academy of Sciences,* 287, pp. 72–85.

SHIVA, V.; BANDOPADHYAY, J. 1980. The Large and Fragile Community of Scientists in India. *Minerva,* 18 (4), pp. 575–94.

TOURAINE, A. 1988. *Return of the Actor: Social Theory in Post-industrial Society.* Minneapolis, Minn., University of Minnesota Press.

VALIANATOS, E. 1976. *Fear in the Countryside: The Control of Agricultural Resources in the Poor Countries by Nonpeasant Elites,* Cambridge, Mass., Ballinger.

VANDERMEER, J. 1981. Agricultural Research and Social Conflict. *Science for the People,* January, pp. 5–8; February, pp. 25–30.

WILKES, G. 1983. Current Status of Crop Germplasm. *Critical Reviews in Plant Sciences,* 1 (2), pp. 131–81.

Biotechnology and economic restructuring: towards a new technological paradigm in agriculture?

G. Otero

Introduction

Modern agriculture finds itself today at a crossroads: it has become economically unsustainable, through the mounting cost of inputs, and environmentally unsound through its intense reliance on petrochemicals. The question arises now as to whether biotechnologies might provide an option to steer the future of agriculture in a sustainable–regenerative direction, while maintaining and increasing its productivity achievements. An alternative development could be, of course, that biotechnologies will prolong the petrochemical era of agriculture, leading rural societies further into social polarization and environmental degradation.

World economic restructuring

The development of biotechnologies coincides with a phase of deep restructuring of world capitalism, covering what has been called a "new technological revolution", led by microelectronics, and including important advances in the area of new materials, informatics, robotics and biotechnologies. One possibility is that this technological revolution will lead to a "reformed capitalism" (Sklair, 1989). The transitional phase involves major transformations in institutional arrangements, in the international division of labour, and in the appropriation and control of knowledge.

INSTITUTIONAL CHANGES

Three major institutional changes associated with world economic restructuring may be highlighted. First, there is a general dismantling of all forms of welfare states, both in capitalist and state-socialist societies, involving not only the abandoning of a number of social functions which the state had been assuming since the 1930s, but also an increasing withdrawal from its directly productive functions, reflected in an increasing privatization of the state-sector in the economy.

A second institutional change concerns the emergence of new industrial relations in which a key concept is flexibility. This trend is clearly associated with the general decline of important sectors of heavy industry (e.g. steel), which were the most dynamic economic sectors at the turn of the century, but are now facing the need for profound restructuring (Piore and Sabel, 1984) – a process which has been described as representing a crisis in the Fordist model of mass production (Kenney and Florida, 1988). This decline of formerly crucial sectors of industry is accompanied by the emergence of the so-called informal sector, in which productive units lie beyond state control, as regards both taxation and labour and wage conditions. In general, industrial relations in the informal sector operate against wage-earners, for their organizational, labour and wage conditions tend to be depressed (Portes *et al.*, 1989). Furthermore, while flexibility may require a more skilled work force and greater worker participation in decision-making in some sectors, its darker side is the sweat shop and home workers (Benería and Roldán, 1987; Rowbotham, 1990). This also leads to a regressive tendency in income distribution within and among countries: poor countries are becoming poorer and rich ones are becoming richer (Durning, 1990).

Finally, there is a tendency for market forces to be re-established as the main allocator of economic resources, and to eliminate subsidies, protectionism and other forms of state intervention in the economy which disturb what the market would otherwise do when left alone. This means that the logic of profit maximizing prevails, with the market forces being in fact controlled by a decreasing number of large transnational corporations (Corbridge, 1990; Green, 1990; Salles, 1990).

NEW INTERNATIONAL DIVISION OF LABOUR

Profound changes in the international division of labour are also taking place. Firstly, the collapse or exhaustion of the import substitution industrialization (ISI) model adopted in the larger countries of Latin America, North Africa and the Near East. The main purpose of this model was to develop a local industry based on local capitalists, using tax incentives, subsidies and protectionism, all of which also involved a strong and direct state intervention in the strategic sectors of the economy, from which important industrial inputs were produced and sold at subsidized prices. One of the expectations of this model was that, in due course, protectionism and subsidies would be eliminated, as local industry developed an international competitiveness. Unfortunately, such competitiveness did not materialize.

On the other hand, the agricultural sector was subordinated to the goal of industrialization in most developing countries which adopted the ISI model. Agriculture was expected to secure foreign exchange - much needed for the import of machinery and other industrial inputs - through the export of commercial crops; to produce cheap food for the growing urban population; and, through its own modernization, to provide a labour force for employment in the industrial sector. In fulfilling these multiple roles, agriculture was led to a major crisis of decapitalization and pauperization of the rural population from which it has not recovered. This is true for both Latin America and Africa (De Janvry, *et al.,* 1989; *The Economist,* 1990). There were net transfers of resources from agriculture to industry for over two decades. Food crops have been left behind, generating a loss of food self-sufficiency in many poor countries. Such a downturn of agriculture explains the growth in foreign indebtedness as a source of foreign exchange to continue financing the industrialization process in Latin America, until indebtedness also reached critical proportions in the 1980s.

Secondly, there is an increasing substitution of raw materials formerly exported by developing countries for new materials now produced in advanced countries. The most recent and dramatic case of substitution, this time by a biotechnology derived product, has been sugar. Several Caribbean countries and the Philippines saw their sugar exports drop heavily. The USA substituted almost half of its previous sugar imports for locally produced high fructose corn syrups - HFCS (Ahmed, 1988). This technological development was also made possible thanks to the protectionist policy of the USA, which imposes very high import tariffs to sugar in order to protect its local producers. Such policy led industrial consumers of sugar in the USA to promote the production of HFCS with enzyme technology (Otero, 1989a). On the other hand, some South American countries are experiencing what has been called a "perverse integration" to the world economy, through the cultivation and processing of illegal crops (Castells and Laserna, 1989). To the extent that traditional export crops from developing countries continue to be substituted by new materials, this type of perverse integration could become more generalized.

Thirdly, the newly industrialized countries (NICs) of Asia (Hong Kong, Singapore, the Republic of Korea and Taiwan) have become important protagonists in the world economic scene, with their industrialization being oriented toward exporting manufactures from the outset, quickly moving into the export of goods based on high technologies; their dependency was based more on foreign aid (e.g. credit) and foreign trade than on direct foreign investment; the bulk of their exports is accounted for by local capitalists and not by transnational corporations as in Latin America (Gereffi, 1989); and, finally, development in these NICs has been more "inclusionary" than that in Latin America, in the sense that the standard of living of their working classes has increased much more substantially.

Fourthly, a reshuffling of economic blocs is taking place, especially in Europe, North and Central America, and the Pacific Rim countries, which would exclude most developing countries, and it is quite likely that new forms of protectionism among blocs will emerge, which could lead to increased economic isolation for developing countries, while the bulk of world trade takes place among advanced capitalist countries (Sewell, *et al.*, 1988).

PRIVATIZATION OF KNOWLEDGE

Scientific knowledge and its technological applications are becoming increasingly subject to private appropriation, as reflected in the proliferation of patents (Berlan, 1989; Otero, 1989b, 1989c), or the growth in secrecy among scientists with industrial research relationships (Kenney, 1986; Curry and Kenney, 1990). See the chapter by Kenney.

Regarding the links between this privatization process and developing countries, a major paradox has emerged regarding world genetic resources, the raw material for continued improvements in crop varieties, whether through conventional breeding and / or biotechnologies. It so happens that the poor countries geographically located in the South have been naturally endowed with the greatest genetic riches, but they lack the capital, scientific and technical resources to develop new plant varieties; conversely, the rich countries of the North, very poorly endowed in plant genetic resources, have had free and indiscriminate access to the resources from the South. Rich countries and their powerful seed industries have profitably used those resources in the production of improved varieties and although they have often returned them to the countries of the South, they have done so in the form of commodities, which must be purchased at a price (Kloppenburg, 1988; Kloppenburg *et al.*, 1988). It remains to be seen whether biotechnologies might represent a way out of poverty for developing countries as 60% of the population in the developing countries lives from agriculture and will be directly or indirectly affected by the coming biorevolution (Ahmed, 1989; Otero, 1991).

Modern agriculture: can this technological paradigm be transcended?

The term "modern agriculture" here refers mainly to a technological package, made up of hybrid and improved plant varieties, petrochemical fertilizers, herbicides, pesticides, mechanization and irrigation, similar to the package exported to developing countries during the "green revolution". This package became the "technological paradigm" (Dosi, 1984) for "modern" agriculture. The crops most affected by this model, in terms of hybrid and improved plant varieties, were maize, rice and wheat, although the rest of the package was extended to a large number of crops which have been subjected of massive applications of chemicals. It was adopted as a complete package, mostly in areas with irrigated agriculture, in Latin America, the Near East and North Africa, while in Asia a selective adoption was made, with much less intensive mechanization. By contrast, few regions in sub-Saharan Africa have adopted the "green revolution" technologies (International Labour Organisation, 1988).

One of the consequences of this differential adoption of technological innovation is that vast regions of the world have become largely marginalized from the world economy (Paarlberg, 1988). Nevertheless, these marginalized regions of the world might have the possibility of moving directly beyond the "petrochemical era" of agriculture into one which relies onbiotic processes to increase production and expand employment opportunities, while at the time preserving economic and environmental sustainability of agriculture.

Unfortunately, the current economic and institutional trends give rise rather to pessimism than to optimism. Biotechnologies will probably deepen the structural changes brought about by the "green revolution" and, given the very different institutional context in which they are emerging, they are likely to have far more socially and regionally polarizing effects than those associated with the "green revolution". The major obstacle to be overcome in order to move in the direction of an alternative agriculture, both in the USA and in developing countries, may be the current structure of inputs producers in the agri-food complex, which consists of oligopolistic transnational corporations which seem to direct biotechnology research in the wrong direction (Golberg *et al.*, 1990).

Biotechnologies, employment and environment

SUBSTITUTIONISM AND EMPLOYMENT

It seems that biotechnology's greatest short- and medium-term threat to employment in developing countries is precisely its substitutionist tendency. A longer term possibility, however, is the creation of totally new products. For this to become an opportunity rather than a threat, future scientific research would have to be geared to addressing the needs of developing countries, with an explicit effort to utilize the raw materials at their disposal, as well as developing scale-neutral technologies or even technologies biased in favour of the small-scale producers. The problem arises, however, of how to steer research in the direction of meeting broader human needs, as it is the transnational corporations which mainly determine the research agendas, clearly defined by their profit motive.

BIOTECHNOLOGIES AND THE ENVIRONMENT

The 1980s witnessed an increase in the environmental concerns raised by modern agriculture, for example: over-dependence on non-renewable resources; soil erosion, degradation and deforestation; pollution of surface and groundwater supplies. Conversely, proponents of an alternative agriculture which would be sustainable base their argument on the fact that there is a growing population which calls for increasingly efficient production; that there is the need to produce energy from biomass, i.e. from agriculture, as hydrocarbons are a non-renewable resource, and nuclear energy raises safety and environmental problems; and that such a system would save the family farm, which represents a personalized form of agricultural practice and a whole way of life (Bidwell, 1986; Francis, 1988; Rodale, 1988).

Other forms of alternative agriculture, such as biological farming and agro-ecology (Altieri, 1987) have also been advocated, notably in a report by the US National Academy of Sciences, entitled *Alternative agriculture* (October, 1989), of which the most important conclusion is that alternative farming practices can achieve both economic and environmental viability.

The potential of biotechnologies for promoting an alternative agriculture did indeed arouse optimism for environmentalists, but one decade later disillusionment has set in. Herbicide-tolerant or pesticide-tolerant varieties, instead of freeing agriculture from petrochemical dependence, may tend to expand the technological package of modern agriculture with its attendant problems. The exchange of herbicide-tolerant genes between the domesticated crops and weedy relatives could ultimately result in the need for more herbicides to control herbicide-resistant weeds (Golburg *et al.*).

The increase in agricultural surface treated with herbicides has been dramatic in the past three decades. In 1956, only 11% of acreage planted to maize were treated with herbicides, whereas in 1988, herbicide use on three major crops – maize, soybeans and cotton – has increased to roughly 95%. It has been estimated that less than 1% of the pesticides, including herbicides, applied actually reach target pests. Consequently, more than 99% of herbicides applied may contaminate land, water, air, humans, other animals, and wildlife habitat. Moreover, herbicides account for approximately 31% of the estimated oncogenic risk of pesticide residues in fresh foods and approximately 12% of the pesticide residue risk in processed food (Golburg *et al.*, 1990).

This situation benefits the agricultural inputs industry. Only 8 pesticide-producing companies now account for about 70% of proprietary pesticide sales world-wide. Those are the same companies that are supporting research on herbicide tolerance. As a consequence of this market structure, farmers do not benefit from the productivity gains which may result from herbicide use, for they face an oligopsonistic market when they sell their commodities to the food and fibre processing industry.

If herbicide-resistant varieties are introduced in developing countries, a further erosion of genetic diversity of crop and wild plants, and exacerbated pesticide-related human health and environmental problems may be expected. Displacing existing varieties could contribute to the extinction of traditional landraces and cultivars.

BIOTECHNOLOGIES AND ALTERNATIVE AGRICULTURE

Biotechnologies can contribute to avoiding the problems created by modern agricultural technologies, while preserving and expanding the productivity gains. It should also be feasible to generate technologies which favour the small scale or are scale-neutral. The major obstacle is the current market structure of both the agricultural inputs and the food and fibre processing industries.

The technical options available or feasible for an alternative agriculture include those recommended in the USA Biotechnology Working Group, namely: enhanced use of natural systems, such as biological nitrogen fixation, in the agricultural production process; reduced use of off-farm inputs, such as pesticides; greater use of existing genetic potential of plants and animals; improved accomodation between cropping patterns, and the climatic and physical limitations of agricultural land; profitable and efficient production emphasizing improved management and conservation of natural resources, such as soil, water, wildlife and wildlife habitats.

While these suggestions may preserve the environment, it is questionable whether they will also contribute to expanding agricultural productivity. For this purpose, a more agressive approach is necessary, aimed at maximizing the positive potential of biotechnologies. For example, *Rhizobium* inoculants may be developed as a substitute for synthetic fertilizers, to help boost productivity without the environmental degradation caused by chemical fertilizers. Microbial pesticides may also be developed as a substitute for chemical pesticides.

One obvious way of promoting this type of beneficial biotechnologies is more government funding for research. Attention should also be given to the ways in which the uses of genetic manipulation may reduce dependence on hazardous chemicals in agricultural production systems and promote long-term sustainability of highly productive agriculture (Goodman *et al.,* 1987).

Conclusions

In order to face the coming revolution of biotechnologies with some degree of success, developing countries must adopt both offensive and defensive strategies. The first type of strategy involves identifying their local scientific and production capabilities, and trying to orient research and development toward a range of goods and processes not currently contemplated by private firms in developed countries. The products and processes created by local scientists and producers should eventually be used for forming new niches in international trade, which might offset some of the losses in jobs and foreign exchange caused by the substitution of some traditional exports of developing countries by new products of biotechnologies. Local biotechnologies should be used to promote an alternative agriculture which would be intensive and productive yet environmentally and socially sustainable.

With regard to defensive strategies, studies should be conducted on two major fronts. It is necessary to assess, firstly, which agricultural exports of developing countries are most endangered by new or potential products of biotechnologies, and secondly, what alternative uses of current agricultural commodities may be profitably promoted, and / or what other land uses should be encouraged in those areas where a certain product has ceased to be profitable.

References

AHMED, I. 1988. The Bio-revolution in Agriculture: Key to Poverty Alleviation in the Third World? *International Labour Review,* Vol. 127, No. 1.

AHMED, I. 1989. Advanced Agricultural Biotechnologies: Some Empirical Findings on their Social Impact. *International Labour Review,* Vol. 128, No. 5.

ALTIERI, M. 1987. *Agroecology: The Scientific Basis of Alternative Agriculture.* Boulder, Westview Press.

BENERIA, L.; ROLDAN M. 1987. *The Crossroads of Class and Gender: Industrial Homework, Subcontracting, and Household Dynamics in Mexico City.* Chicago, University of Chicago Press.

BERLAN, J.P. 1989. The Commodification of Life. *Monthly Review,* Vol. 41, No. 7, pp. 24–30.

BIDWELL, O.W. 1986. Where Do we Stand on Sustainable Agriculture? *Journal of Soil and Water Conservation,* September–October 1986.

CASTELLS, M.; LASERNA, R. 1989. The New Dependency: Technological Change and Socio-economic Restructuring in Latin America. *Sociological Forum,* Vol. 4, No. 4.

CORBRIDGE, S. 1990. Post-Marxism and Development Studies: Beyond the Impasse. *World Development,* Vol. 18, No. 5, pp. 623–9.

CURRY, J.; KENNEY, M. 1990. Biotechnology in Land-grant Universities. *Rural Sociology,* Vol. 55, No. 1.

DE JANVRY, A.; SADOULET, E.; YOUNG, L.W. 1989. Land and Labour in Latin American Agriculture from the 1950s to the 1980s. *Journal of Peasant Studies,* Vol. 16, No. 3, pp. 396–424.

DOSI, G. 1984. *Technical Change and Industrial Transformation.* London, Macmillan.

DURNING, A. 1990. Ending Poverty. In: Brown, L.R. *et al. State of the World 1990: A Worldwatch Institute Report on Progress toward a Sustainable Society.* New York, London, Norton.

FRANCIS, C.A. 1988. Sustainable Versus Non-sustainable Resources: Options for Tomorrow's Agriculture. *National Forum,* Summer 1988.

GEREFFI, G. 1989. Rethinking Development Theory: Insights from East Asia and Latin America. *Sociological Forum,* Vol. 4, No. 4.

GOLBURG, R.; RISSLER, J.; SHAND, H.; HASSEBROOK, C. 1990. *Biotechnology's Bitter Harvest: Herbicide-tolerant Crops and the Threat to Sustainable Agriculture.* A report of the Biotechnology Working Group (USA).

GOODMAN, R. *et al.* 1987. Gene Transfer in Crop Improvement. *Science,* 236, pp. 48–54.

GREEN, R.H. 1990. La evolución de la economía internacional y la estrategia de las transnacionales alimentarias. *Comercio Exterior,* Vol. 40, No. 2, pp. 91–100.

INTERNATIONAL LABOUR ORGANISATION (ILO). 1988. *Rural Employment Promotion.* Geneva, International Labour Conference, 75th Session, Report VII.
KENNEY, M. 1986. *Biotechnology: The University-industrial Complex,* New Haven, Yale University Press.
——. ; FLORIDA, R. 1988. Beyond Mass Production: Production and the Labor Process in Japan. *Politics and Society,* 16 (1), pp. 121–58.
KLOPPENBURG, J. JR. 1988. *First the Seed: The Political Economy of Plant Biotechnology.* New York, Cambridge University Press.
——. ; KLEINMAN, D.; OTERO, G. 1988. La biotecnología en los Estados Unidos y el Tercer Mundo. *Revista Mexicana de Sociología, No. 1.*
OTERO, G. *1989a. Industry–university Relations and Biotechnology in the Sugar and Dairy Industries: Contrasts between Mexico and the United States. In: World Employment Programme Research Working Paper.* Geneva, International Labour Organisation (ILO).
——. 1989b. *Commodification of Science: Biotechnology in the United States and Mexico.* Paper presented in the meetings of the Society for Social Studies of Science, (Irvine, Cal., USA, 15–18 November 1989).
——. 1989c. Ciencia, nuevas tecnologías y universidades. *Ciencia y Desarrollo,* Vol. XV, No. 87, pp. 49–59.
——. 1991. The Coming Revolution of Biotechnology: A Critique of Buttel. *Sociological Forum,* Vol. 6, No. 2 (forthcoming).
PAARLBERG, R.L. 1988. U.S. Agriculture and the Developing World: Opportunities for Joint Gains. In: Sewell, J.; Tuker, S.K. *et al. Growth Exports and Jobs in a Changing World Economy: Agenda 1988,* pp. 119–38. New Brunswick (USA), Oxford (UK), Transaction Books.
PIORE, M.J.; SABEL, C.F. 1984. *The Second Industrial Divide: Possibilities for Prosperity.* New York, Basic Books.
PORTES, A.; CASTELLS, M.; BENTON, L.A. 1989. The Informal Economy: Studies in Advanced and Developing Countries. Baltimore, London, Johns Hopkins University Press.
RODALE, R. 1988. Agricultural Systems: The Importance of Sustainability. *National Forum,* Summer 1988.
ROWBOTHAM, S. 1990. Postfordism. *Z. Magazine,* September 1990.
SALLES, S. 1990. *Las posibilidades limitadas de las biotecnologías.* Paper presented in the VII Latin American and Caribbean Seminar for Food Science and Technology, (San Jos), Costa Rica, 2–7 April 1990).
SEWELL, J.W.; TUKER, S.K. *et al.* 1988. *Growth, Exports and Jobs in a Changing World Economy: Agenda 1988.* New Brunswick (USA), Oxford (UK), Transaction Books.
SKLAIR, L. 1989. *Assembling for Development. The Maquila Industry in Mexico and the United States.* Boston, Unwin Hyman.
The Economist (London). 1990. Africa's Cities. 15–21 September 1990, pp. 25–8.

Biotechnologies végétales : évolution et implications

S. S. Filho

Introduction

Les biotechnologies sont de plus en plus incorporées aux stratégies concurrentielles des grandes firmes de la chimie, de la pharmacie, des industries alimentaires et des semences. En conséquence, les relations entre le secteur public et le secteur privé deviennent plus étroites, parce que ce dernier veut avoir accès à la connaissance engendrée dans les institutions publiques et souhaite voir une part importante des risques de l'investissement en recherche biotechnologique rester à la charge du secteur public. Toutefois, les longs délais nécessaires à la rentabilisation de ces investissements et la grande incertitude relative aux possibilités de commercialisation font que le développement des biotechnologies se déroule encore largement dans le secteur public. Cela est particulièrement vrai pour les biotechnologies végétales.

Biotechnologies végétales : caractéristiques générales

Le secteur agro-alimentaire est moins attrayant, à court terme, pour les investissements en biotechnologies que le secteur pharmaceutique. Ses produits ont, en général, une valeur ajoutée moindre ; ils proviennent d'organismes biologiquement plus complexes et posent des problèmes de propriété industrielle plus difficiles à résoudre (Paillotin, 1988). Le problème principal réside toutefois dans les stratégies concurrentielles des entreprises industrielles.

Par exemple, dans le domaine de la résistance aux herbicides, il est beaucoup moins coûteux pour les entreprises impliquées d'adapter une plante à un herbicide que de mettre au point un nouveau produit. Les coûts pour sélectionner une nouvelle variété résistante sont estimés à $ 2 millions environ, tandis qu'un nouvel herbicide demande un investissement de l'ordre de $ 40 millions. Ce n'est donc pas un hasard si parmi les 10 compagnies qui contrôlent le marché mondial des semences, 8 sont aussi impliquées dans celui des herbicides, et si les 10 plus grandes compagnies d'agrochimie interviennent sur le marché des semences. La valeur de ce dernier est estimée à $ 3 milliards vers le milieu des années 90 et à $ 6 milliards à la fin du siècle (GEBM, 1988). Cela indique qu'en dépit des difficultés d'ordre technico-scientifique relatives au développement de la biologie moléculaire des plantes, les stratégies des grands groupes industriels déterminent l'évolution future des biotechnologies végétales. Cela paraît d'autant plus vrai que la sélection de plantes résistantes aux herbicides, en tant que stratégie concurrentielle, n'aura pas seulement recours à des techniques intermédiaires, mais aussi au génie génétique. C'est ainsi que la compagnie Calgene Inc. a obtenu des plants de tabac transgéniques, résistants à l'herbicide Bromoxinyl, à la suite de l'introduction d'un gène chimérique qui permet à la plante de synthétiser une enzyme qui neutralise l'action phytotoxique du produit (*Biofutur,* janvier 1990). La société belge Plant Genetic Systems a aussi réussi à transférer les gènes de résistance à la phosphinotricine chez la betterave sucrière, la tomate, la pomme de terre et le tabac (*Biofutur,* janvier 1990). Monsanto Co. (USA) a obtenu, par génie génétique, des plants de cotonnier et de tomate résistants à son herbicide « Round up ». La commercialisation de ces variétés transgéniques est prévue vers le milieu des années 90 (GEBM, 1989). Voir tableau 1.

Dans le domaine de la résistance aux insecticides, des recherches sont faites pour trouver des produits de remplacement ; d'où l'intérêt de la sélection de plantes rendues résistantes, par génie génétique, aux insectes, mais aussi à des micro-organismes pathogènes. Une cinquantaine de firmes paraissent attirées par ce domaine : de grandes compagnies comme Abbot Laboratories, Sandoz A.G., Shell, Monsanto Co. et ICI Ltd., et de nouvelles entreprises de biotechnologie (NEBs) comme Ecogen et Mycogen. Le marché est estimé à quelque $ 20 ou 30 millions et, pour le moment, les dépenses de recherche-développement (R&D) sont supérieures aux ventes à cause des longs délais d'aboutissement des projets en cours. C'est à cause de ces risques économiques et de ces

Tableau 1. Compagnies et institutions effectuant des recherches sur la résistance des plantes aux herbicides

Compagnie	Contractant	Herbicide	Culture
Allelix		Atrazine	Colza
ARCO (PCRI)	Heinz	Atrazine	Tomate
Biotechnica Inc.		Atrazine	Soja
Calgene Inc.		Perenmedipham	
Calgene Inc.		Glyphosate	Cotonnier et maïs
Calgene Inc.	Rhône-Poulenc	Bromoxinyl	Tournesol
Calgene Inc.	Kemira-Oy.	Glyphosate	Colza
Calgene Inc.	Nestlé	Atrazine	Soja
Calgene Inc.	Campbell's	Glyphosate	Tomate
Calgene Inc.	Dekalb-Pfizer	Glyphosate	Maïs
Calgene Inc.	Coker's Seed Co.	Glyphosate	Tabac
Calgene Inc.	Phytogen	Glyphosate	Cotonnier
DuPont		Chlorosulfurone	Tabac
DuPont		Sulfometurone	
Mobay (Bayer)		Metribuzine	Soja
Molecular Genetics	American Cyanamid	Imidazolinone	Maïs
Monsanto Co.		Glyphosate	
Phyto-Dynamics		Trifluraline	Maïs
Shell		Atrazine	Maïs
Cornell University		Triazines	Maïs
Harvard University		Atrazine	Soja
Louisania State University		Glyphosate	
Michigan State University		Atrazine	Soja
Rutgers University		Triazines	
University of Alabama		Atrazine	
University of California (Davis)		Sulfometurone	Tournesol
University of Guelph (Ontario, Canada)		Atrazine	Colza
U.S. Department of Agriculture (Agricultural Research Service)		Metrubuzine	Soja
U.S. Forest Service		Glyphosate	Espèces forestières
Ciba-Geigy AG		Atrazine	
Plant Genetic Systems	Hoechst A.G.	Basta	

Source: Kloppenburg J. R. 1988. *First the seed.* Cambridge, Cambridge University Press.

délais que ce type de recherche reçoit plus d'attention sur le plan institutionnel. Au Japon, par exemple, le Ministère de l'industrie et du commerce (MITI) met au point un projet de recherche d'une durée de

10 ans sur les phéromones et autres messagers chimiques, en y investissant quelque $ 65 millions (*Biofutur,* septembre 1989).

Les orientations de la recherche relatives à la résistance aux herbicides et au contrôle biologique des maladies ne sont en rien contradictoires du point de vue des stratégies concurrentielles : en effet, les entreprises cherchent d'un côté à stimuler des marchés très prometteurs, celui des herbicides, et, de l'autre, à agir dans des secteurs qui offrent déjà des perspectives technologiques pour entretenir ou créer des marchés.

La production de semences artificielles constitue un autre domaine intéressant pour les stratégies concurrentielles. L'isolement et la mise en capsule des embryons somatiques peuvent aussi apporter de profonds changements dans le marché des semences. On peut, en effet, espérer étendre l'utilisation des variétés hybrides à d'autres cultures que le maïs, car de tels embryons pourraient résulter de fusions interspécifiques ; d'autre part, le problème de la propriété industrielle se trouve résolu en raison de la protection naturelle des hybrides.

Mais ces semences artificielles peuvent être accompagnées d'engrais, de fongicides ou d'herbicides. Cela est déjà utilisé pour la tomate et la carotte aux USA et pourrait concerner des céréales à l'avenir ; il s'agit de véritables « kits » de productivité qui pourraient conduire à une nouvelle vigueur des marchés traditionnels d'engrais et de pesticides, comme dans le cas des recherches sur la résistance aux herbicides.

Le génie génétique appliqué aux végétaux n'est qu'à ses débuts et n'a pas encore d'applications commerciales. Les difficultés d'obtention des plantes transgéniques sont grandes, surtout pour les céréales. Des variétés transgéniques sélectionnées pour augmenter les rendements doivent permettre des gains de productivité élevés (6 % à 8 %) pour que les coûts de R&D puissent être compensés (Baccon-Gibody, 1988). Les résultats les plus intéressants concernent la résistance aux herbicides et la production de plantes mâles stériles.

Le tableau 2 récapitule les contrats de R&D et / ou de commercialisation dans le domaine agro-alimentaire aux USA, en Europe et au Japon, entre décembre 1988 et novembre 1989. Ce tableau, élaboré à partir d'un rapport spécial paru dans la revue *Biofutur,* montre que le secteur agro-alimentaire se trouve au deuxième rang des secteurs d'application des biotechnologies, après celui de la santé humaine, où le pourcentage total de contrats est de 61 % contre 27 % dans le secteur agro-alimentaire. D'autre part, une partie significative de ces contrats concerne des opérations d'achat et / ou de fusion d'entreprises, surtout de semences et d'additifs alimentaires ; cela montre que la concentration économique qui a commencé dans ce secteur dans la première moitié des années 80, se poursuit. Cela indique aussi que la R&D biotechnologique est influencée de façon nette par les stratégies concurrentielles des grands groupes industriels concernés.

Tableau 2. Contrats de recherche-développement (R&D) ou de commercialisation de produits issus des biotechnologies dans le domaine agro-alimentaire (décembre 1988 à novembre 1989)

Agriculture (22 contrats ; 14 % du total)	– 8 contrats d'entreprises semencières. – 5 contrats de R&D et / ou de commercialisation de variétés résistantes aux maladies. – 2 contrats de R&D et de commercialisation de diagnostics. – 1 contrat de R&D sur les marqueurs génétiques. – 3 contrats de commercialisation de biopesticides. – 1 contrat de R&D sur la résistance aux herbicides. – 2 fusions d'entreprises de biopesticides.
Élevage (6 contrats ; 4 % du total)	– 5 contrats de R&D et / ou de commercialisation de vaccins. – 1 projet d'aquiculture.
Industrie alimentaire (14 contrats ; 9 % du total)	– 7 fusions d'entreprises alimentaires et additifs avec des NEBs ou des entreprises établies. – 3 contrats de R&D pour de nouveaux micro-organismes fermentaires. – 2 contrats de commercialisation de cyclodextrines. – 2 contrats pour la mise au point de nouveaux additifs.
	N. B. Dans le secteur de la santé humaine, on a dénombré 98 contrats (61 % du total) ; dans celui de l'environnement, 4 contrats (2 % du total) ; et dans divers secteurs, 16 contrats (10 % du total). Le nombre total des contrats est de 160.

Source : Biofutur. (Paris), février 1990.

Implications du processus

Une biotechnologie traditionnelle, intermédiaire ou de pointe, peut avoir plusieurs applications, parfois complètement distinctes les unes des autres. Il en est ainsi à cause de simples empêchements technique ou juridique (propriété intellectuelle), mais aussi à cause de la logique concurrentielle des secteurs impliqués ou des stratégies des entreprises et des industries.

Dans les années 80, l'entrée de grands groupes de la chimie, de la pharmacie, d'aliments et de semences, indiquait que les biotechnologies constituaient une option importante de diversification. Au cours de cette période, les NEBs ont cherché une fin honorable à leurs premières prétentions non satisfaites ; aujourd'hui, elles sont, sauf rares exceptions, des entreprises de service ou qui fabriquent certains produits spécialisés. En 1988, le bilan réalisé pour 20 NEBs a montré un sensible progrès par rapport à 1987, et que 4 entreprises connues faisaient des profits : Genentech Inc., Amgen, Applied Bioscience et Genzyme ; les chiffres d'affaires de mai à août 1988 ont été de 44 % supérieurs à ceux de la même période de 1987. De toute façon, la plupart des investissements réalisés pendant les années 80 n'auront des résultats qu'à partir de 1990. Le capital risque en biotechnologies s'est aussi accru de 2 % aux USA en 1988 par rapport à 1987 (*Biofutur,* janvier 1990). En ce qui concerne les domaines d'action, le secteur des biotechnologies végétales progresse moins vite et se trouve systématiquement plus faible que ceux de la santé humaine et de la médecine vétérinaire.

En ce qui concerne les facteurs qui influent sur les stratégies concurrentielles, le premier d'entre eux se rapporte aux limitations technico-scientifiques imposées aux procédés de genèse et de diffusion de l'innovation technologique. Isoler, transférer et faire exprimer les gènes d'un micro-organisme représentent des opérations moins complexes que celles qui portent sur une cellule végétale ou sur une cellule animale. C'est pourquoi le succès commercial du génie génétique a été pour le moment confiné à la recombinaison chez des micro-organismes.

Le deuxième facteur se rapporte aux questions de propriété industrielle en biotechnologies. Sur le plan juridico-économique, on constate la victoire des entreprises américaines dans les débats sur la brevetabilité de procédés et produits, qui s'étend même au règne animal. Le problème, c'est-à-dire en dernière instance celui de la propriété intellectuelle, influe de façon déterminante sur l'évolution récente des biotechnologies. Deux raisons au moins expliquent cette affirmation : l'imperfection du système de brevets dans le monde et les possibilités de contourner l'originalité des organismes susceptibles d'être brevetés. Dans le dernier Rapport de l'OCDE, on note à ce propos qu'« ...il n'existe aucun autre domaine technologique où les lois diffèrent sur tant de points et divergent de manière si importante entre les pays que celui de la biotechnologie » (OCDE, 1989). D'autre part, il est difficile de définir avec exactitude les caractéristiques génétiques (dans le cas des organismes) et chimiques (dans le cas des molécules) ; de s'assurer que le produit mis au point ne soit pas identique à un autre produit naturel ; et enfin d'avoir des garanties que ne seront pas obtenus des organismes ou des molécules à peine modifiés, mais ayant les mêmes fonctions. A titre d'exemple, on peut citer le cas du facteur VIII de la coagulation sanguine, utilisé par les hémophiles et produit par génie génétique par

Genentech Inc., dont la demande de brevetage a été refusée en raison de l'existence d'un brevet antérieur sur le facteur VIII extrait du sang ; on peut aussi citer le cas du procès, également perdu par Genentech Inc., au sujet de la protropine (déjà brevetée) et la humotropine de Eli Lilly & Co. (brevetée ensuite), qui sont toutes deux des hormones de croissance humaines obtenues par génie génétique ; elles diffèrent par un seul acide aminé et cela a paru suffisant pour les considérer comme des médicaments originaux du point de vue de la propriété industrielle.

Le troisième facteur qui concerne la nature et le degré de participation du secteur public, se trouve impliqué dans le développement de la structure de la R&D, le type et le degré d'interaction avec le secteur privé, le pouvoir de réglementation, de normalisation et de fiscalisation, ainsi que dans la participation directe au secteur productif.

L'expansion des marchés de produits issus des biotechnologies nécessite une structure de R&D un peu institutionnalisée, c'est-à-dire qu'il faut compter sur des institutions établies, mais aussi sur un organisme responsable de la politique scientifique et technologique, c'est-à-dire de la définition des priorités et de l'orientation du développement. Quant à l'articulation directe entre les universités ou instituts de recherche publics et le secteur productif, les universités américaines sont « instrumentalisées » pour servir directement le secteur privé, en particulier depuis la promulgation de la loi 96-517 de 1982 qui a contraint les universités à rechercher des ressources en dehors de la sphère publique et à breveter leurs découvertes (Kloppenburg *et al.*, 1983). L'industrie est devenue, dans le domaine des sciences biologiques, la principale source de financement de la recherche académique (Orsenigo, 1989) ; voir tableau 3. Il est aussi important de souligner la

Tableau 3. Investissements industriels dans les universités

Groupe industriel	Année	Institution	$ millions / années
Hoescht A.G.	1981	Massachusetts Hospital	70 / 10
W. R. Grace	1981	Massachusetts Institute of Technology	30 / 5
DuPont	1982	Harvard Medical School	6 / 5
Exxon	-	Cold Spring Harbour Laboratory	7,5 / 5
Monsanto Co.	1982	Washington University	23,5 / 5
	1987	Washington University	40 / 5
Corning Glass	-	Cornell University	2,5 / 6
Kodak	-	Cornell University	2,5 / 6
Union Carbide	-	Cornell University	2,5 / 6

Source : Ducos, C.; Joly, P. B. 1988. *Les biotechnologies.*

participation de scientifiques dans les entreprises, surtout par l'intermédiaire des comités scientifiques consultatifs (Kloppenburg, 1988).

Les dispositions ou les mesures prises en matière de réglementation, de normalisation et de fiscalisation modifient les coûts, le temps de commercialisation et les décisions d'investissement et la réalisation des recherches. C'est le cas, par exemple, des normes techniques établies pour les aliments, les additifs et les médicaments, ou des instruments de politique industrielle (soutien financier, propriété industrielle et encouragement à la R&D et à l'importation).

Quant à la participation directe de l'État dans la production, elle peut modifier l'évolution des biotechnologies, dans la mesure où l'État peut intervenir dans des domaines jugés prioritaires sur le plan social, par exemple pour mettre au point des systèmes d'immunothérapie, produire des denrées alimentaires de base ou lutter contre la pollution.

En résumé, l'évolution des biotechnologies végétales est déterminée par les stratégies concurrentielles des entreprises, qui sont, à leur tour, influencées par trois types de facteurs : des limitations technico-scientifiques, des questions liées à la propriété intellectuelle et les formes institutionnelles d'action de l'Etat. Cela a les implications suivantes :

a. les produits et les procédés ont pour fonction première de répondre aux priorités de la concurrence et du marché ; la consolidation de la première étape commerciale des biotechnologies dans le secteur agro-alimentaire s'appuiera sur la récupération ou le renforcement des marchés traditionnels (intrants chimiques, semences et plants) ;

b. les produits et procédés seront de préférence ceux qui ont une protection naturelle (nouveaux hybrides) ou qui n'ont pas besoin de protection, mais qui peuvent être facilement diffusés (variétés résistantes aux herbicides et variétés ayant des propriétés importantes pour le traitement agro-industriel, comme, par exemple, une plus grande teneur de substances solides solubles dans les tomates et les oranges) ;

c. les limites technico-scientifiques constituent un obstacle au développement des biotechnologies, dans la mesure où elles exigent un soutien financier important des activités de R&D, qui n'est pas à la portée des petites et moyennes entreprises ;

d. le soutien institutionnel joue un rôle important dans la définition des activités de R&D, car une infrastructure forte en recherche fondamentale contribue beaucoup aux applications technologiques ; les formes d'articulation entre le secteur public et le secteur privé provoquent des conflits dans les deux sphères, qui influent sur le développement de la R&D ;

e. même si une grande partie des recherches sur les végétaux est incorporée aux travaux des entreprises, la dépendance vis-à-vis de la recherche publique est encore très grande ; on observe même le retour de certains travaux de R&D au secteur public, en raison des difficultés financières des entreprises.

Incidences pour les pays latino-américains

L'orientation des biotechnologies par les grandes compagnies agro-alimentaires ainsi que la concentration et la centralisation des capitaux dans les secteurs de l'agrochimie, des semences et des industries alimentaires ont pour effet d'aggraver la dépendance technologique en général, de réduire les avantages commerciaux existants et de conduire à un système agricole qui a un effet d'exclusion encore plus marqué que l'actuel modèle de production. Sur ce dernier point, une étude récente réalisée par le Ministère de la coopération d'Allemagne a conclu que 74 % des petits cultivateurs ne pourraient pas profiter des nouvelles technologies et seraient exclus du système productif. Seule une petite partie des grandes exploitations agricoles pourra bénéficier des biotechnologies (*Biofutur,* décembre 1989).

Toutefois, en s'efforçant de surmonter les limitations technico-scientifiques définies au préalable, en agissant sur la législation de la propriété intellectuelle, en favorisant l'intégration entre les secteurs de production et les institutions de recherche, les pays latino-américains pourraient modifier certains impacts négatifs et occuper alors leur « place au soleil ». On peut citer à cet égard les activités de R&D suivantes : obtention de nouvelles variétés et de nouveaux hybrides pour des cultures régionales, grâce aux techniques de culture de tissus qui sont en grande partie maîtrisées par les pays latino-américains ; isolement à partir de cultures de tissus ou de cellules de métabolites secondaires d'usage pharmaceutique ou alimentaire ; amélioration de la productivité animale par les techniques de manipulation d'embryons et par la mise au point de vaccins ; lutte biologique contre les maladies et fixation biologique de l'azote atmosphérique ; utilisation des tests à base d'anticorps monoclonaux pour détecter des maladies de plantes (Lamptey & Moo-Young, 1987). Ces exemples n'excluent pas des techniques plus sophistiquées comme celles du génie génétique.

Mais l'établissement de politiques et de priorités de recherche dépend aussi des solutions apportées à l'appropriation des ressources génétiques (Kloppenburg *et al.,* 1988). De surcroît, cette attitude offensive se fonde sur la capacité du secteur public en recherche fondamentale et appliquée, sur l'établissement de normes qui permettent d'atteindre les objectifs proposés et sur des mécanismes plus dynamiques d'articulation avec le secteur privé. Comme il n'y a pas dans les pays latino-américains suffisamment d'industries novatrices, il revient au secteur public de jouer un rôle central dans la compétitivité interne et externe.

Le cas du Brésil

Au Brésil, un premier exemple d'illustration des questions évoquées précédemment se rapporte à la fermeture d'une des plus grandes entreprises nationales de biotechnologies végétales, Biomatrix, due à des difficultés financières (élévation des coûts de l'entreprise et étroitesse des marchés auxquels elle destinait ses produits). Le second exemple est celui de Bioplanta qui est la deuxième entreprise de biotechnologies végétales liée à Companhia Souza Cruz, qui appartient à British American Tobacco Co. ; cette compagnie a aussi des difficultés dues aux coûts très élevés des recherches et à l'étroitesse des marchés ; cette entreprise a pendant 3 ans investi de 33 % à 50 % de son chiffre d'affaires dans la R&D, ce qui s'est avéré insoutenable même pour une entreprise bénéficiant du soutien d'une grande compagnie multinationale (Bonacelli, 1989).

Le troisième exemple est celui du Centro Nacional de Recursos Genéticos da Empresa Brasileira de Pesquisa Agropecuria (CENARGEN–EMBRAPA), institution publique de recherche où se trouve la principale équipe de génie génétique du pays et dont les projets (parmi lesquels on trouve le transfert au haricot des gènes impliqués dans la synthèse de la méthionine dans la « castanha do Par ») sont largement soutenus par des fonds publics. Le quatrième exemple est celui du Projet Bio–Rio, qui est un parc de biotechnologies en voie d'établissement à Rio-de-Janeiro : 70 entreprises en 10 ans. Jusqu'en 1990, $ 7 millions ont été investis à fonds perdus ; les terrains pour l'installation des entreprises appartiennent à l'Université fédérale de Rio-de-Janeiro ; des facilités de crédit et des avantages fiscaux sont accordés aux entreprises, mais une grande partie des produits sera achetée par le gouvernement.

On peut se demander si les investissements en biotechnologies dans les pays latino-américains peuvent continuer de dépendre presque exclusivement de fonds publics et, si le secteur privé doit jouer un rôle important, avec quels capitaux et quelles entreprises il pourra mener cette tâche.

Références

BACCON-GIBODY, J. 1988. Colloque d'Angers : de l'*in vitro* à la production de plantes et à l'amélioration génétique. *Biofutur* (Paris), janvier 1988.
BIOFUTUR. 1989. Bioactualité (82), septembre 1989.
——. 1989. Vie des sociétés (83), octobre 1989.
——. 1989. Instantanés (85), décembre 1989.
——. 1990. Bioactualité (86), janvier 1990.
Bio / technology. 1989. vol. 7, n° 3.
——. 1990. vol. 8, n° 3.
BONACELLI, M. B. H. 1989. *Processo de capacitaçao tecnològica em biotecnologia : um estudo de caso.* NPCT / UNICAMP, 31 pp.
DUCOS, C. ; JOLY, P. B. 1988. *Les biotechnologies.* Paris, Éditions La Découverte, 128 pp.
GEBM (GENETIC ENGINEERING AND BIOTECHNOLOGY MONITOR, UNIDO). 1988. (24).
——. 1989. (26).
——. 1990. (27).
KLOPPENBURG, J. R.JR ; KLEINMAN, D. L. ; OTERO, G. 1988. La biotecnología en Estados Unidos y el Tercer Mundo. *Revista Mexicana de Sociología,* 2 (1).
——. 1988. *First the seed.* Cambridge, Cambridge University Press, 349 pp.
LAMPTEY, J. ; MOO-YOUNG, M. 1987. *Biotechnology : Principles and options for developing countries.* In : Da Silva, E. J. ; Dommergues, Y. R. ; Nyns, E. J. ; Ratledge, C. 1987. *Microbial technology in developing world.* Oxford, Oxford University Press. 444 pp.
OCDE. 1989. *Biotechnologie. Effets économiques et autres répercussions.* Paris, OCDE, 128 pp.
ORSENIGO, L. 1989. *The emergence of biotechnology ; institutions and markets in industrial innovation.* London, Pinter Publishers, 220 pp.
PAILLOTIN, G. 1988. *L'avenir des biotechnologies dans l'agriculture et l'agro-alimentaire.* NPCT-UNICAMP, mimeo.
SALLES FILHO, S. L. M. ; BONACELLI, M. B. M. ; DEL BIANCHI, V. 1987. *Biotecnologia e produçao de alimentos.* NPCT / UNICAMP, 84 pp.

Industrial biotechnology policies: guidelines for developing countries

F.C. Sercovich

Introduction

The transition from biotechnology research to biotechnology manufacturing is not as easy as is sometimes assumed. Save a few exceptions, like that of *in-vitro* diagnostic kits, the passage from the laboratory to the industrial arena is less trivial than many enthusiasts admit. Furthermore, entry into biotechnology as an industrial activity is not purely firm-specific, but involves a whole series of interacting agents.

In developing countries, the accumulation of basic biotechnological knowledge does not easily filter through to the economic sphere. For a workable transition to be made from scientific effort to the market, a wide variety of capabilities and institutions have to be in place, such as a reasonably well-articulated risk capital market; an enterprise sector permeable to scientific culture, and vice versa; and corresponding institutions and legal codes.

In developing countries, much depends on what standards are set to define entry into biotechnologies. They cannot afford to take false steps by adhering to loose guidelines. This refers not only to scientific quality but also to industrial, engineering, organizational and entrepreneurial standards, which can in no way be met unless due attention is paid to gaps in technological mastery, polyvalent engineering skills and scale-up related issues.

Market entry

To date, most biotechnological developments are sharply at odds with views that suggest that biotechnologies are particularly suitable for developing countries, because of what they promise, their allegedly low entry barriers and appropriateness for leap-frogging. However, basic techniques are being routinized, the technological trajectory is becoming increasingly user-specific and imitation costs and time lags are being shortened, all of which may facilitate entry by developing countries, provided that scaling-up and downstream processing problems are addressed appropriately. Sometimes the push from science is stronger than the pull coming from industry or vice versa, while strong market-driven elements can be identified in some cases.

Cuba is a good example of a science-driven entry into largely health-oriented biotechnologies, mainly at the research-and-development stage. Although some production capacity was developed, it cannot reach world markets because of allegedly deficient quality standards (so far Cuba is only serving some Third World markets based on concessionary assistance and science and technology co-operation deals). Its cost competitiveness is unknown. The Centre for Biological Research (CIB), set up in 1982, produces its own restriction enzymes and carries out research on the synthesis of oligonucleotides, the cloning and expression of a number of other genes (e.g. those encoding interferons), and the production of monoclonal antibodies (Mabs) for diagnostic purposes. Cuba's entry into biotechnologies pursued social ends: the interest in interferon was prompted by the outbreak of dengue haemorrhagic fever affecting some 300,000 people in the late 1980s. There was also, however, a science-push drive: first-rate biologists were available and it was thought that biotechnologies suited Cuba because of its research-intensive nature. If Cuba is to take steps to get closer to the world market, substantive efforts will have to be made to set up cost-efficient and world quality process, product, and production engineering standards, as well as marketing and distribution channels.

Argentina's entry into biotechnologies shows strong industry-push elements. There are a few biotechnology firms working in the field of diagnostics, vaccines and micropropagation, led by two small pioneer firms mainly active in human health. The predicament facing one of these firms is typical of the situation in developing countries (i.e. external diseconomies and expensive in-house efforts). Having mastered recombinant DNA and related techniques for the purpose of an initially modest project, the firm found itself obliged to extend the scope of the project in order to produce a wider range of products. Apparent shortcuts drawing on imitation and extensive use of freely available information turned into unexpected bottlenecks, requiring much unforeseen experimental work which resulted in a 6-year delay in commercial production and a substantial budget increase (Katz and Bercovich, 1988). Although the project was technically feasible, its economic rationale remains to be demonstrated. No industrial policy framework was available to support this effort.

Much stronger and effective demand-pull elements are found in Brazil. The elements behind the rationale for the Alcohol Programme were energy dependency, and an expected price of a barrel of petroleum over $ 40. Brazil's headstart in the field of ethanol from cane sugar relied on natural advantages and on the mastery of all capabilities needed to turn out complete package deals, including project design, execution and start-up, process know-how, machinery construction, training, technical assistance and planning of integrated agro-industrial operations. The Programme relied largely on known fermentation engineering, scaling-up and mass production, rather than on genetic engineering. However, the Programme, along with the exploitation of a variety of biomass sources, created a large and avid market for biotechnology breakthroughs (Sercovich, 1986). This is a good illustration of the role of synergistic factors in biotechnological development, but only a few developing countries can afford such a comprehensive effort.

Brazil's headstart in conventional biotechnologies has spun off an incipient development of more advanced biotechnologies. University-industry links are being forged through initiatives like Bio-Rio, a science park that will offer an incubator facility, central laboratories for sequencing and synthesis of nucleotides, genetic engineering experiments and scale-up, administrative support and technical services.

While in Latin America the weak link is usually industry, in developing South-East Asian countries it is the domestic science base. The Republic of Korea, Singapore, Taiwan and Thailand show comparatively stronger market-driven orientations. They also have industrial policies which are more explicitly focused on biotechnologies, including supply of credits, grants, risk capital, and support for training and process and product development. Thailand pays more attention to agricultural applications and the other countries to health-related applications. In Singapore, Taiwan and Thailand start-ups play an important role. The Republic of Korea relies much on large conglomerates that devote substantial resources to biotechnology research-and-development. South-East Asian countries offset the relative weakness of their science base by drawing directly on industrialized countries' scientific establishment through their expatriates, and by setting up biotechnology research firms there. Samsung and Lucky–Goldstar (Republic of Korea) have done so in the USA (Yuan, 1988). Conversely, Glaxo is setting up a $50 million research joint venture with the Institute of Molecular and Cellular Biology (IMCB), Singapore. (*Genetic Engineering News,* 1989, p. 26).

In Africa, Zimbabwe takes advantage of expatriate scientists working in France in the area of DNA probes for Salmonella. This work is of global interest as the disease causes 3.5 million deaths each year in children (*The Economist,* 1990, p. 81).

Joint research ventures do not necessarily work to the advantage of developing countries. Some agreements may enable firms based in industrialized countries to use developing countries' research skills and capabilities as a source of cheap inventive labour whose output is subsequently processed industrially and commercially back in the industrialized countries (Thayer, 1989, p. 7; *Chemical and Engineering News,* 1989, p. 14). The Chinese are involved in this kind of joint–research venture while acquiring, at the same time, turn–key, pre–fabricated biotechnology facilities from a major multinational in order to manufacture recombinant hepatitis B vaccines (*The Wall Street Journal,* 1989).

Industrial policy issues

The science–push drive fails to work in some cases, e.g. vaccines, where price competition apparently discourages leading firms from engaging in development and manufacturing. As long as technological and manufacturing barriers are not overcome, a number of vaccines that can be produced today on the basis of existing scientific knowledge simply will not reach those who need them. Because industrialized countries' markets do not justify their commercial development, they remain expensive and therefore beyond the reach of those who need them most.

Developing countries remain relatively backward, despite their potential for catching up, because they lack many or all of the ingredients for forming the social capability required to realize such potential. There should be no illusions as to biotechnologies being an exception in this regard. Many developing countries can endow a group of first–rate scientists, at the cost of great sacrifices, with the resources necessary to undertake high–quality research, but to expect to be able to reach the world market on this basis is an illusion. As Japan, and then the Republic of Korea, Hong Kong, Singapore and Taiwan have shown, the key to effectively exploiting the leap–frogging potential does not simply lie in the mastery of the scientific underpinnings of a technology, but rather, in that of the engineering, industrial and commercial capabilities that make it possible to reach the market competitively. Although less successful, Brazil and Mexico have been trying to apply the same lesson.

The intensity of current international competitive rivalry and the fact that the USA, the leading country in the field, is trying to offset its declining competitiveness, is an unfortunate coincidence for developing countries. Conditions for access to technological know-how are harder than they used to be, when much knowledge and information regarding manufacturing processes was transferred on a commercial basis. Today such transfers to developing countries are rare, and the rapidly shifting scientific, technological and industrial frontiers in biotechnologies accentuate the uncertainties. For instance, initial price quotations for biotechnology products are very high, since the firms concerned intend to recover research-and-development costs as quickly as possible, yet prices may go down substantially at any time, which makes it difficult for developing country firms to make a realistic assessment of future returns.

Although developing countries may have little chance of entering directly into high value-added product lines involving heavy research and development expenses, they do have certain indirect strategic routes for taking effective economic and social advantage of biotechnologies. It would be illusory, however, to attempt to commercialize biotechnology products without paying attention to the mastery of effective downstream processing technologies through joint work between chemical engineers and biochemists. The lack of bioprocess engineering skills may effectively block scale-up efforts, particularly at the purification stage. The ability to undertake effective scale-up is a major entry barrier into most commercial biotechnology segments. Genetic engineering has permitted mass production of proteins and lower costs for products such as enzymes and amino-acids, but it does not substitute for more conventional engineering disciplines.

Many biotechnology start-ups based in industrialized countries are eager to engage in technology transfer agreements with firms based in developing countries. However, it is necessary to proceed with caution since, in most cases, their technologies are still at an experimental stage. On the other hand, there are also many instances of successful applications of the outputs of such research in developing countries, e.g. Zimbabwe's DNA probes for Salmonella, Argentina's diagnostic test for Chagas disease and Colombia's malaria vaccines (Eisner, 1988).

Conclusions

One of the main problems facing developing countries with regard to biotechnologies is that of knowing how to enter the market at the right time and how to avoid dead ends. The dynamics of biotechnological change in developing countries must be understood in order to identify technology and market trends, and the different actors: universities, research units, dedicated biotechnology firms, or multinational corporations. It is also essential for developing countries to understand the nature of the most important factors that affect the timing of introduction and rate of diffusion of biotechnologies, such as company strategies, scientific, technological and engineering bottlenecks and uncertainties, barriers to entry and threshold factors, and the relative competitiveness of biotechnological products and processes. Considerable time and resources will be required to bridge the gap between the rapid development of the scientific basis and the lagging evolution of the technological and manufacturing processes.

The inability of developing countries to supply products and services at competitive prices downgrades their ability to generate wealth. High value-added products make it possible to pass on high costs of research, but for the moment they do not appear to be the solution for developing countries attempting to enter the biotechnology market.

Once the basic biotechnology techniques become routinized, one of the main questions is what to do with them. The answer to this question can only result from a learning process whereby the accumulation of scientific, technological and manufacturing skills and capabilities interacts with social needs and market realities. This process entails, on the one hand, the carrying out of basic and applied research on a continuous basis and, on the other, the setting-up of the engineering capability that is needed to translate the resulting insights into competititve products. This is precisely what the Japanese appear to have understood very early. The first stage (1981–1988) consisted of achieving mastery of the basic biotechnology techniques by taking full advantage of research links with the best centres of excellence in the world. The second stage (1988 onwards) consists of acquiring the necessary manufacturing experience through licences and then starting to enter the market as innovators, forging ahead at the scientific, technological and commercial levels (Masuda, 1989).

International technical co-operation includes support for the setting-up of information networks; the strengthening of domestic scientific and technological capabilities (e.g. bioprocess skill formation, experimental development and scale-up, setting-up and upgrading standards of manufacturing, quality, and process and product safety, working out of industrial policy guidelines); assistance in the transfer and adaptation of technology; support for the development of new products and processes. Initiatives such as the Programme of policy research and technical assistance in biotechnology (PRATAB) (Sercovich and Leopold, 1990) would help in tackling an urgent need to avoid duplication, create synergies and improve the use of resources. PRATAB is intended to perform as a scanning and early warning system for the benefit of developing countries through technical assistance and policy research in biotechnologies. A network of data banks would be set up and consulting services would be provided to governments and organizations in developing countries.

References

Chemical and Engineering News, 3 April 1989. Cell Technology in Chinese Joint Venture.

EISNER, R. 1988. *Genetic Engineering News.* (July / August 1988).

Genetic Engineering News, June 1989.

KATZ, J.; BERCOVICH, N. 1988. *Biotecnologia e industria farmaceutica.* Buenos Aires. United Nations Economic Commission for Latin America and the Caribbean (ECLAC).

MASUDA, M. 1989. Bio-industry policy reviewed in mid-term. *Business Japan,* July 1989.

SERCOVICH, F. 1986. The Political Economy of Biomass in Brazil – The Case of Ethanol. In: Jacobsson *et al. The biotechnical challenge.* Cambridge, United Kingdom, Cambridge University Press.

——. ; LEOPOLD, M. 1990. *Developing Countries and the "New" Biotechnology: Market Entry and Related Industrial Policy Issues.* Ottawa, International Development Research Centre (IDRC), Manuscript Series.

THAYER, A.M. 1989. US firm, Chinese Set up Biotech Venture. *Chemical and Engineering News,* 20 March 1989.

The Economist. 13 January 1990. The Slow March of Technology, p. 81.

The Wall Street Journal, 1989.

YUAN, R.T. 1988. *Biotechnology in South Korea, Singapore and Taiwan.* Washington, D.C., International Trade Administration.

Biotechnology and rural labour absorption

I. Ahmed

Introduction

Agriculture is the primary source of income of the world's poor, who live almost entirely in rural areas and whose livelihood depends on agriculture, whether or not their income is derived directly from it (World Bank, 1990). In general, the rural poor belong to wage labour or marginal farmer households; their poverty results as much from low returns as from unemployment and underemployment. The numbers of the rural poor have increased from 767 million in 1970 to 850 million in 1985 (Singh and Tabatabai, 1990).

On a global basis between the early 1960s and the 1980s world food crop production grew a half percent faster than the growth in population (Mellor, 1988), yet the absolute number of undernourished people in developing countries actually increased from 460 million in 1969–1971 to 512 million in 1983–1985 (Singh and Tabatabai, 1990), poverty resulting from the inability to purchase sufficient quantities of food from available food supplies. It is obvious therefore that biotechnology applications could make a contribution to poverty alleviation if they could boost purchasing power by improving labour absorption in rural areas without sacrificing growth in agricultural output.

Impact of biotechnologies on productivity and rural labour absorption

GAINS FROM BIOTECHNOLOGIES

China presents illustrations of breakthroughs made in the application of the advanced plant biotechnologies to two major cereal crops, rice and wheat (Table 2); economic returns from savings in costs of pesticides and gains in productivity are remarkable (Tables 1 and 2). The adoption of biotechnologies in potato cultivation in Kenya doubles land productivity, labour-intensity and profitability and increases labour productivity by 24% (Table 7); its contribution (value added) to national income is twice that of conventional technologies; the relative efficiency (value added as a proportion of gross output) is also higher for the adopters.

Table 1. China: output and profitability of biofertilizer applications in selected crops (1987).

Biofertilizer	Crop	Area (ha)	Increase in yield	Increase in revenue (million yuan)
Root-combined nitrogen-fixation bacteria	wheat	150,000	76.2 million kg	51 (1)
Nodule bacteria	groundnuts	60,000	340 kg / hectare	10
Nodule bacteria	lucerne	200,000	20 %	45
Fungus (gemma fungus)	rice, wheat corn, etc.	100,000	10–30%	
Actinomycetes	grain crops, tobacco leaf, etc.	60,000	40–50 kg / ha for grain (2)	

(1) Revenue increased by 15.7 yuan per 1 yuan investment in biofertilizers.
(2) Average increase of pesticide cost amounted to 1.8–6 yuan per hectare.

Source: Yuanliang (1989)

Table 2.China: output gains from cell-engineering on major crops.

Crop	Type of cell engineering	Crop variety	Sown area (ha) in 1988	Increase in yield
Wheat	chromosomal eng.	nos 4, 5, 6 Xiao Yan	2,000,000	900,000 tonnes
Wheat	culture of pollen (haploid cells)	no. 1 Jing Hua	70,000	15–20 %
Rice	pollen (haploid cells)	Xin Xiu, Wan Gen 959,etc.	170,000 (1)	about 10 %
Rice	pollen (haploid cells)	nos 8, 9 Zhong Hua	70,000 (1)	15–20 %
Rice	marker rescue	no. 1 Hu Yu	3,000 (1)	15 %
Potato	tissue culture		70,000 (2)	over 50 %
Sugar–cane	tissue culture		4,000	over 50 %
Tobacco	pollen (haploid cells)		10,000	over 50 %
Banana	tissue culture		100,000 test–tube seedlings	over 50 %

(1) 1985.
(2) 1984.

Source: Yuanliang (1989).

LABOUR ABSORPTION IN CROP PRODUCTION

Evidence from several International Labour Organisation (ILO) country case studies clearly suggests that with the application of biotechnologies there is a saving in labour–use for chemical means of plant protection. The case study for Mexico shows that the application of micropropagation techniques need not lead to labour displacement in citrus cultivation as this would be compensated by more intensive labour–use in weeding, pruning, irrigation and harvesting from a reduction in crop losses (labour accounted for 78%–82% of total costs of citrus production). In fact, in Malawi and Kenya, there was a substantial increase in labour–use per unit of land following the application of biotechnologies (through the introduction of new practices) which also increased yields. In Kenya it was due to more labour needed for ridging before cultivating potato and in Malawi for nursery and planting operations (Tables 3, 4, 5, 6 and 7).

As a result of the application of plant biotechnologies to a range of crops, a high potential has been noted for an indirect and steady source of seasonal employment through the strengthening of the forward

Table 3.Citrus fruit production and employment in the Puuc region of Yucatan, Mexico, 1987.

Type of employment	Number of jobs	% of total jobs
Total producers	5,246	62.4
Of which:		
small producers (less than 3 ha)	5,123	(97.7)
medium producers (3–10 ha)	100	(1.9)
large producers (over 10 ha)	23	(0.4)
Farm overseers	12	0.1
Agricultural day labourers	3,052	36.3
Employees in juice plant	21	0.2
Lorry drivers	11	0.1
Market vendors	34	0.4
Carriers	30	0.4
TOTAL	8,406	100.0

Source: Eastmond and Robert, 1989.

Table 4.Changes in class structure in the citrus fruit zone of Mexico, 1970 and 1980.

		Percentage of total occupations	
Occupation	Class	1970	1980
Public officials, managers, administrators, large land-owners	upper	1.4	0.3
Professionals and technical workers, administrative personnel, traders	middle	16.2	4.7
Personal service workers, transport workers, small agricultural producers,lower cattlemen, lumbermen, fishermen, hunters, agricultural day labourers, non-agricultural workers, machine operators	lower	82.4	78.0
Unspecified and unemployed	lower	–	17.0
TOTAL		100.00	100.00

Source: Eastmond and Robert (1989).

Table 5. Malawi: percentage change in labour input (work-days) per hectare in different operations by level of technology

Operation	Between Indian hybrid and polyclonal tea	Between polyclonal and clonal tea	Between Indian hybrid and clonal tea
Nursery / new planting	+ 18.3	+ 6.8	+ 25.1
Pruning	+ 15.2	− 15.4	+ 0.1
Weeding	+ 6.8	+ 2.8	− 4.5
Plant protection	− 1.9	− 0.0	− 1.9
Plucking	− 9.6	+ 7.1	− 2.5
All operations	+ 15.2	+ 0.8	+ 16.3

Source: Chipeta and Mhango (1988).

Table 6. Malawi: percentage distribution of labour input (work-days) per hectare by operation and level of technology

	Level of technology		
Operation	Indian hybrid	Polyclonal	Clonal
Nursery / new planting	0.0	21.2	28.3
Pruning	51.7	60.3	42.1
Weeding	17.4	6.4	9.0
Plant protection	4.2	1.3	1.3
Plucking	26.7	10.8	18.8
TOTAL	100.00	100.00	100.00

Source: Chipeta and Mhango (1988).

linkages to the juice processing plant (Mexico), poultry production (Nigeria), coffee, henequen, tequila and dairy industries (Mexico), and the tea industry (Kenya). Underemployment (e.g. 75% – 90%) of the agricultural workers in the south-eastern region of Mexico, caused by seasonality of agricultural production, can be reduced by applying plant biotechnologies to create crop varieties and widen their range, with a view to prolonging the growing season and supplying ripe oranges; this not only cuts down agricultural underemployment but also reduces excess capacity in juice processing (e.g. plant idle for 6 months).

Table 7. Biotechnologies and farm size: potato and tea in Kenya 1987.

Key indicators	Potato farms (N = 33) Biotechnologies (BT)	Potato farms (N = 33) Traditional technologies	Tea farms / estates (N = 39) Relationship with farm size	Tea farms / estates (N = 39) Relationship with farm size biotech. only
Labour productivity (gross output/hectare in shillings)	33,210	16,382	Inverse for BT and TT	Inverse
Labour intensity (work-days/hectare)	301	144	BT: unclear	Unclear
Labour productivity (kg/work-day)	124	100	Positive for both BT and TT	Positive (sh/work-day)
Labour's factor share (wages as % of value added)	27	23	Positive for both BT and TT	Positive
Capital use				
sh/work-day	3	3	Inverse for	Positive
sh/hectare	867	426	BT and TT	Positive
Intermediate inputs (sh/hectare)	3,553	3,008	Inverse for both BT and TT	Inverse
Value added as a proportion of gross output (%)	89	82	BT: positive TT: inverse	Positive
Profitability (gross output minus operating costs in sh/ha)	20,816	9,916	Inverse for both BT and TT	Positive
Income ratio (1)	8	4	-	3(small farms) (2) 2(large estates) (3)

1. Ratio of income of the 30 % of richer farmers to income of the 70 % of poorer farmers.
2. Up to 3 hectares.
3. Over 20 hectares.

Source: calculated from data in Mureithi and Makau (1989).

STRUCTURAL COMPOSITION OF RURAL EMPLOYMENT

There are structural changes in rural and agricultural employment associated with the application of plant biotechnologies which, in China, for instance, releases labour from agriculture which is absorbed in new and side-line activities in specialized occupations, with a change in social organization of the delivery of these services. In Malawi and Kenya, it has led to new farm practices involving increased labour-use (Tables 5 and 6).

In rural labour markets, the application of plant biotechnologies increased the demand for hired labour (e.g. in Mexico for citrus and in Kenya for tea and potato), boosted wages and reduced rural–urban wage differentials. Gross earnings for these technologies in Kenya compare favourably with wage incomes in a modern sector job important for slowing down the pace of rural–urban flow of income–seekers.

Increased herbicide applications have had consequences for women's employment. Despite the availability of chemical herbicides, the 'green revolution' relied heavily on manual labour for weeding, which is one of the most labour-intensive of all agricultural operations. The resultant increase in the demand for hired labour in weeding doubled in Sri Lanka, (Hameed *et al.,* 1977), while overall labour use in weeding doubled or tripled in Bangladesh (Ahmed, 1981) and the Philippines (Bartsch, 1977). Small farmers recorded much higher labour intensity in weeding than larger farmers (Ahmed, 1981) and women constituted between 72% and 82% of the labour used for weeding (Unnevehar and Stanford, 1985). As the introduction of genetically engineered plant varieties will lead to a substitution of chemical herbicides for manual weeding, a massive displacement of women's labour is to be expected. The trends indicate that not only will the genetically engineered biotechnology plant varieties introduce a new fixed cost for farmers by forcing them to purchase the herbicide genetically tied to the seed supplied by the same company, but that it will also strike a major blow at the poor.

BLENDING OF SKILLS

Case studies concerning micropropagation of crops, e.g. in Mexico, Kenya and China, demonstrate the blending of workers with "low-tech" skills to engage in traditional agricultural work, with "highly-skilled" technicians directly involved with biotechnologies. Similarly, in Nepal, scientific personnel can produce 8,000 to 10,000 potato plantlets per day through the application of micropropagation techniques which can then be easily rooted in sandbeds by semi-skilled workers (Rajbhandari, 1988).

Another interesting feature in the Philippines (Halos, 1989) and Mexico (Eastmond *et al.,* 1989) is the predominance of women in the micropropagation laboratories. Women constitute 80%, 74% and 85% of the Philippine Society for Microbiology, Cell and Molecular Biology, and Biotechnology Societies, respectively. In those countries, these were regarded as low-paid jobs concerned with basic sciences with previously limited linkage to industry.

New prospects for multiple cropping

The most important factor which contributed to greater labour use per hectare as a result of the 'green revolution' was the practice of multiple cropping facilitated by the early maturing varieties of cereals. The application of micropropagation techniques to potato could similarly help improve cropping intensity. Since potatoes take only 40–90 days to grow in most developing country climates (compared to 150 days in the temperate climates), this can easily be incorporated into the cropping patterns currently practised for cereals like wheat, rice and corn. Thirty poor countries already have the capacity to micropropagate potatoes. Indeed, micropagation techniques have made potato the second largest crop (by weight) after rice in Vietnam, and have quadrupled the production in China over the past 30 years (*The Economist,* 13 October 1990, pp. 113–4). In Viet Nam, micropropagation techniques have increased potato yields from 200 tonnes to 8,000 tonnes per year on 450 hectares of land within a period of four years (1980–1984) (Uyen and Zaag, 1985). These techniques have already brought about yield increases from 8 tonnes to 18 tonnes per hectare in Nepal (Rajbhandari, 1988). Micropropagation techniques for potato are attractive for employment creation and poverty alleviation for the following reasons: year-round production of plantlets; reduction of cost and difficulty of physical transportation of potato tubers to the fields for planting; direct regeneration of plantlets from plant tissues, with a substantial volume of tubers spared from planting being used as food; and production of disease-free planting material (Manandhar *et al.,* 1988).

Prospects for small farmers

As observed in the case of the 'green revolution', the large farmers pioneer the adoption of plant biotechnologies in Kenya. However, economic inducements exist for all categories of farms for the dissemination of technologies. For instance, the Chinese experience demonstrates the profitability of this diffusion. In Nigeria (Table 8), the high and escalating costs of plant sources of animal feed and the lower relative price of single-cell proteins could serve as an incentive. More than half the small-scale growers of citrus in Mexico were willing to adopt disease-free planting material derived from plant biotechnologies. Those technologies are likely to be more scale-neutral than mechanical or 'green revolution' technologies. Although biotechnological innovations would constitute variable costs, collecting information on which the decision to adopt the technology is based, actually represents a fixed cost. This would constitute an important reason for observing a bias in favour of large farmers (Kinnucan *et al.*, 1989).

Table 8. Nigeria: propensity to substitute single-cell proteins for traditional protein sources (assuming single-cell protein costs 10 % less).

Amount of substitution (%)	Respondents		
	No.	%	Cumulative %
100	13	54.2	54.2
61–99	4	16.6	70.8
41–60	3	12.5	83.3
21–40	3	12.5	95.8
11–20	0	0.0	95.8
5–10	1	4.2	100.0
TOTAL	24	100.0	

Source: Okereke (1988).

Responding to specific socio-economic needs

The orange leaf rust disease ravaging coffee cultivation in Mexico has threatened the survival of the overwhelming majority of small growers who cannot afford chemical means of control. Application of plant biotechnologies to supply disease-free or disease-resistant plant materials will not only save but expand the employment opportunities of these producers and of the large body of hired labour; it will generate indirect employment through its forward linkage to the coffee industry and backward linkages to micropropagation laboratories and nurseries which produce the plantlets.

Plant biotechnologies could be deployed to meet the acute shortage of plantlets for henequen cultivation in Mexico (on which nearly a quarter of Yucatan State's labour force is dependent for its livelihood). Similarly, *in–vitro* tissue culture to provide disease-free planting material for coconut (lethal yellowing affecting a large acreage of plantations annually and to make available a larger number of plantlets at greater speed would provide an important supplementary source of employment to the poor of Yucatan.

Virus–free citrus scions would substantially reduce crop losses (currently 50%) due to pests and diseases, at a relatively low cost, thus enhancing growers' incomes, providing more work for the day labourers, reducing excess capacity in the juice–processing plant and stimulating the local economy of this impoverished region of Mexico.

The tequila agave used to produce the drink tequila is grown in the Mexican state of Jalisco. No less than 12 million plantlets are required to replenish existing stocks. Micro–propagation techniques again offer the answer. Some 6,000 small contract growers who supply the large agro–industrial companies with the raw material will stand to benefit (Eastmond *et al.*, 1989). Through the forward linkages to the tequila industry, additional and more stable employment will be generated.

In Mexico, the use of bovine somatotropin hormone (BST) could reduce the daily deficit of 12.5 million litres of milk by increasing milk production in dairy cows by 10–25%. BST hormone also offers prospects for Pakistan, which has to spend about $ 30 million importing mostly dried milk each year (*The Economist,* 13 January 1990). Similarly, the use of single–cell proteins in the poultry feed industry in Nigeria could reduce acute protein malnutrition and generate employment directly and indirectly.

Plant genetic engineering and job creation

From the point of view of labour absorption and poverty alleviation, the following observations can be made on the current or foreseen plant genetic engineering breakthroughs:

- pest and disease resistance, and drought tolerance will reduce output variance, which is important for risk– averse farmers; together with the breakthroughs for nitrogen fixation these will obviously reduce resource–poor farmers' costs of production;
- production of drought tolerant crop species or varieties will increase labour absorption through area expansion and multiple cropping now made feasible;
- lower labour requirement in pest and disease control may be compensated for by overall increases in labour use in other new operations;
  genetically engineered herbicide resistance will directly displace labour for weeding;
- prolonging the shelf life of freshly harvested agricultural produce will help

the poor confronted with inadequate marketing infrastructure (communication constraints and lack of preservation or refrigeration facilities);
- genetically engineered microbes may benefit the small farmers if these spill over to the poor neighbours' plots and fix nitrogen there or protect the crops from pests and diseases;
- the major obstacles to access by developing countries and poor farmers to beneficial biotechnologies are the legal and financial barriers associated with the proprietary rights over these technologies through patents.

Immediate employment potentials

In-vitro micropropagation, followed by clonal multiplication of crop species or varieties is within the scientific and financial reach of developing countries. It has been estimated that a fully equipped laboratory excluding land and buildings might cost $ 250,000 (Lipton and Longhurst, 1989). Micropropagation techniques can be immediately deployed to enhance rural employment. The capacity to generate micropropagated disease-free planting material in Mexico (agave) and Nepal (potato) has already been demonstrated to be cheaper than the imported planting material. Indeed, this technique is already applied to potato in 30 poor countries (*The Economist,* 13 October 1990). Singapore has the capacity (Plantek International Ltd) to provide disease-free coffee plantlets for large-scale plantation throughout South-East Asia. Similarly, the government agronomic research agency in Brazil (EMBRAPA) is able to produce coffee plantlets (*Biotechnology Development Monitor,* 1990). This scientific capacity of developing countries is being channelled to cover mostly non-food crops and to meet the needs of the commercial and large farm sector, although these may create employment indirectly for hired rural workers.

The Biotechnology Department of the Indian Ministry of Science and Technology has supported (University of Delhi) micropropagation techniques for bamboo (40,000 plantlets by 1989–1990), oil-palm (5,000 plantlets by January 1990), coconut and rubber. Some commercialization has been initiated or achieved for cardamom and bamboo (Mani, 1990). Similarly in the Philippines, out of the 28 commercial and semi-commercial tissue culture laboratories owned privately or by the government, 22 are devoted solely to orchid propagation. Only five laboratories propagate food and fibre crops, and two laboratories which are solely devoted to food crops are both government-owned.

Similarly in Mexico, the micropropagation enterprises are geared to market plantlets for nonfood crops. Certainly, maize, beans and other cereals which constitute the basic Mexican diet receive little attention.

Maize is grown primarily by the peasants and the application of micropropagation techniques to maize is not profitable under current prices.

Coping with external shocks

Developing countries annually lose $ 10 billion of their exports due to the biotechnology-based product substitutions (Kumar, 1988) with serious repercussions on the international division of labour. For instance, while 90% of the sugar internationally traded came from the developing countries in 1975, it declined to about 67% in 1981 (see the chapter by Otero). Sugar imports by industrialized countries declined from 70% to 57% during the same period. World consumption of high fructose corn syrups (HFCS), which accounted for only 1% of total sweeteners in 1975, rose to 6% in 1985 (Wald, 1989). Thirty-four different soft-drink manufacturers in the USA have switched to the HFCS. As a direct consequence, sugar exports from the Philippines declined from $ 624 million in 1980 to $ 246 million in 1984. In the Caribbean the decline of sugar export to the USA was of nearly the same magnitude during this period. This was accompanied by a crash in sugar prices from US cents 63.20 per kg in 1980 to US cents 8.36 in 1985. It is hardly surprising that the livelihood of over 50 million workers engaged in the sugar industry, mostly in developing countries, were affected by the decline in their exports (Panchamukhi and Kumar, 1988). See the chapter by Junne.

Similar tendencies to replace vanilla flavour by biotechnology substitutes threaten 70,000 small farmers in Madagascar which could also lose US $ 50 million of its annual export earnings (Mushita, 1989). Cocoa, the second most important agricultural commodity in developing countries, faces a similar threat of substitution by biotechnology products, with regard to cocoa butter (Hobbelink, 1989).It is reported that current biotechnology research in Germany is oriented to produce a substitute for coffee (Otero, 1990). Developing countries almost exclusively account for the world coffee exports. Apart from the adverse impact on the balance of payments of the major coffee exporters (Brazil, Colombia, Burundi, Uganda, Rwanda and Ethiopia), the livelihood and jobs of 500,000 small producers (average size 1 hectare) in Rwanda and another 650,000 in Indonesia would be threatened (*Biotechnology Development Monitor,* 1989) if the above substitution efforts succeed.Apart from the job losses associated with the changes in the North–South trade flows, biotechnology applications to convert vegetable oils to produce structural lipids or tailored fats will affect the market shares of 11 oilseed species traded by the developing countries (Kumar, 1988). For instance, while coconut is the source of only 2% of the world's oils and fats market, the Philippines alone supplies 80% of the coconut. Decline in Philippine coconut exports will directly affect the 15 million Filipinos dependent on the coconut industry for their livelihood, the majority of whom are poorer than the rest of the farming population (Halos, 1989).

There is, therefore, an urgent need for internal adjustment of production structures of the affected countries to redeploy the workers, particularly from the plantations and small-farm sectors made redundant by the decline in the international demand for their exports. In order to minimize the adverse effects of such external shocks, it is necessary to be vigilant about new biotechnological developments related to other important export crops, so that appropriate structural adjustment measures can be adopted in good time.

Conclusions

RURAL LABOUR MARKETS

Biotechnologies, on the one hand, displace labour for pest and disease control and, on the other, increase labour use on account of new operations and increased cropping intensity or area expansion. Impact on employment (different categories of workers) depends on existing labour market structures. It is important to examine what opportunities biotechnologies offer for wage labour in countries without well-developed labour markets. While plant biotechnologies increase the demand for hired labour, it is often the adult labour which is hired. Even if it increases the income of a typically landless agricultural labour family, the non-availability of the adult male member of the household places a greater work burden on women for unpaid household work. In this respect, it would resemble the effect of migration by male family members.

It is very important to identify the category of labour displaced by biotechnologies. If they displace primarily family labour, the impact on the market for hired labour will be limited. On the other hand, if biotechnology applications lead to the massive displacement of hired labour, this will have a significant depressing effect on the rural wage rates for the related agricultural operation. The workers displaced in one agricultural operation may not possess the skills to undertake the tasks newly created by biotechnologies.

TECHNOLOGY DIFFUSION

There is a need to identify specific policies which enable the small farmers to adopt the new biotechnologies at about the same time as the large farmers. The question of a more equitable land ownership distribution remains important in future agendas for agrarian reform from the point of view of growth, efficiency, labour absorption and equity, both in a traditional agricultural setting and under dynamic conditions of technological change. In this connection, it is to be noted that the high labour absorption capacity of agriculture in East Asian countries, such as China, the Republic of Korea and Taiwan, was primarily due to the egalitarian distribution of land, usually associated with a low incidence of hired labour (International Labour Organisation, Rural Employment Promotion, 1988).

STRUCTURAL ADJUSTMENT, GREATER VIGILANCE AND MEETING SOCIO-ECONOMIC NEEDS

In countries threatened by massive unemployment brought about by a decline in their major agricultural exports on account of biotechnology-induced product substitutions in the industrialized countries, it is necessary to assess the need for the restructuring of the economies of the affected countries, so as to redeploy and retrain the workers made redundant, as, for example, in the developing countries, which are critically dependent on vegetable oil exports.

Research should be directed to identifying policies and measures which will promote development of biotechnologies which respond to specific socio-economic needs, particularly those of the poor in developing countries. Regarding non-farm employment, additional research is needed to quantify the impact of advanced plant biotechnologies on aggregate rural employment through better understanding of the mechanisms and magnitude of backward and forward linkages, and to examine the prospects of applying bovine-somatotropin hormone, single-cell proteins and transgenic pasture crops to the animal husbandry sector if rural non-farm activities are to be expanded.

Many developing countries possess the relatively cheaper scientific capacity regarding micropropagation, but it needs to be strengthened. Their work priorities would obviously have to be re-focused on labour-intensive food crops for mass consumption. Moreover, such facilities could enhance total employment by blending the sophisticated skills of high-level scientific professionals, primary women, with the traditional agricultural labour force.

ACCESS TO BENEFICIAL TRANSGENIC PLANTS AND MICROBES

If the poor cannot afford to buy the biotechnology-derived seeds, the same adverse scenario as was observed for the 'green revolution' will be repeated. Moreover, lack of knowledge on the intricacies of the laws of intellectual property rights and lack of adequate staff greatly restrict developing countries' bargaining power at formal North–South negotiating forums.

References

AHMED, I. 1981. *Technological Change and Agrarian Structure: a Study of Bangladesh.* Geneva, International Labour Organization (ILO).

AHMED, I. 1988. The Bio-revolution in Agriculture: Key to Poverty Alleviation in the Third World? *International Labour Review* (Geneva), Vol. 127, No. 1.

——. 1989. Advanced Agricultural Biotechnologies: Some Empirical Findings on their Social Impact. *International Labour Review* (Geneva), Vol. 128, No. 5, September–October.

BARTSCH, W.H. 1977. *Employment and Technology Choice in Asian Agriculture.* New York, Praeger.

Biotechnology Development Monitor (The Hague), 1989, 1990.

Chicago Tribune, (Wednesday 22 August 1990).

CHIPETA, C.; MHANGO, M.W. 1988. *Biotechnology and Labour Absorption in Malawi Agriculture.* World Employment Research Working Paper WEP 2-22 / WP. 191. Geneva, ILO.

EASTMOND, A.; ROBERT, M. 1989. *Advanced Plant Biotechnology in Mexico: a Hope for the Neglected?* World Employment Research Working Paper WEP 2-22 / WP.200. Geneva, ILO.

EASTMOND, A.; GONZALEZ, R.L.; SALDANA, H.L.; ROBERT, M.L. 1989. *Towards the Application and Commercialisation of Plant Biotechnology in Mexico.* Unpublished draft. Universidad Autonoma de Yucatan, Mexico.

HALOS, S.C. 1989. *Biotechnology Trends: A Threat to Philippine Agriculture?* World Employment Research Working Paper WEP 2-22 / WP.193. Geneva, ILO.

HAMEED, N.D.A. ET AL. 1977. *Rice Revolution in Sri Lanka. Geneva, United Nations Research Institute for Social Development (UNRISD).*

HOBBELINK, H. *1989. Agricultural Biotechnology and the Third World: International Context, Impact and Policy Options. Development-related Research: The Role of the Netherlands.* Groningen, University of Groningen.

International Herald Tribune (25–26 August 1990).

INTERNATIONAL LABOUR ORGANIZATION (ILO). 1988. *Rural Employment Promotion.* Geneva. Report VII, International Labour Conference, 75th Session.

——. (ILO). 1988. *The Challenge of Employment: Rural Labour, Poverty and the ILO.* Geneva

——. (ILO). 1990. *Structural Adjustment and its Socio-economic Effects in Rural Areas.* Geneva. Advisory Committee on Rural Development, Eleventh Session.

KINNUCAN, H.; MOLNAR, J.J.; HATCH, U. Theories of Technical Change in Agriculture with Implications for Biotechnologies. In: Molnar,J.; Kinnucan, H.(eds.). *Biotechnology and the New Agricultural Revolution.* Boulder, Colorado, Westview Press.

KUMAR, N. 1988. Biotechnology Revolution and the Third World: An Overview. In: *Bio-*

technology Revolution and the Third World. New Delhi, Research and Information System for the Non-Aligned and other Developing Countries.

LIPTON, M.A.; LONGHURST, R. 1989. *New Seeds and Poor People.* London, Unwin Hyman.

MANANDHAR, A.; RAJBHANDARI, S.; JOSHI, P.; RAJBHANDARI, S.B. 1988. Micropropagation of Potato Cultivars and their Field of Performance. In: *Proceedings of national conference on science and technology.* Kathmandu, Royal Nepal Academy of Science and Technology.

MANI, S. 1990. Biotechnology Research in India: Implications for Indian Public Sector Enterprises. *Economic and Political Weekly* (Bombay), 25 August 1990.

MELLOR, J.W. 1988. Global Food Security Strategies: Evolution and Role. *World Development,* Vol. 16, No. 9.

MUREITHI, L.P.; MAKAU, B.F. 1989. *Biotechnology and Farm Size in Kenya.* World Employment Research Working Paper WEP 2-22 / WP. 194. Geneva, ILO.

MUSHITA, A.T. 1989. The Impact of Biotechnology in Developing Countries. *Development* (Society for International Development, Rome), 2 / 3.

OKEREKE, G.U. 1988. *Biotechnology to Combat Malnutrition in Nigeria.* World Employment Research Working Paper WEP 2-22 / WP. 190. Geneva, ILO.

OTERO, G. 1990. *The Impact of Biotechnology.* Geneva, ILO.

PANCHAMUKHI, V.R.; KUMAR, N. 1988. Impact on Commodity Exports. In: *Biotechnology Revolution and the Third World.* New Delhi, Research and Information System for the Non-Aligned and other Developing Countries.

RAJBHANDARI, S.B. 1988. Plant Tissue Culture Method and its Potential. In *Proceedings of National Conference on Science and Technology.* Kathmandu, Royal Nepal Academy of Science and Technology.

SINGH, A.; TABATABAI, H. 1990. Facing the Crisis: Third World Agriculture in the 1980s. *International Labour Review,* Vol. 129, No. 4. Geneva.

The Economist. 1990. Let the Sky Rain Potatoes. London, 13 October.

——. 1990. The slow march of technology, London, 13 January.

UNNEVEHAR, L.J.; STANFORD, M.L. 1985. Technology and the Demand for Women's Labour in Asian Rice Farming. In: *Women in Rice Farming.* Aldershot, Gower for International Rice Research Institute (IRRI).

UYEN, N.V.; ZAAG, P.V. 1985. Potato Production Using Tissue Culture in Vietnam: The Status after Four Years. *American Potato Journal,* Vol. 62.

WALD, S. 1989. The Biotechnological Revolution. *OECD Observer,* (Organisation for Economic Co-operation and Development). Paris, 156.

WORLD BANK. *World Development Report 1990.* Washington, D.C.

YUANLIANG, MA. 1989. *Modern Plant Biotechnology and Structure of Rural Employment in China.* Geneva, ILO.

The impact of biotechnology on international trade

G. Junne

Introduction

Biotechnologies are expected to have a major impact on world trade, because they increase the scope for the substitution of one commodity for another. The present situation, however, may differ significantly from historical experience in that switches to a new raw material base may actually take place more quickly than in the past; a large number of commodities will undergo major changes in supply and demand simultaneously, and alternative sources for foreign exchange earnings may be more limited at present (Junne, 1987a; Junne *et al.*, 1989).

Substitution processes

In the area of trade in agricultural commodities, in which the impact of biotechnologies will be felt most, the following four major types of substitution process may be identified.

SHIFTS AS A RESULT OF NEW PLANT CHARACTERISTICS

Genetic engineering of plants has proved much more difficult than that of micro-organisms, but more traditional biotechnologies have already had a considerable impact on plant breeding, especially through tissue and plant cell culture. This has helped to speed up conventional plant breeding and to reduce the lead time for developing new plant varieties. The ease with which new characteristics can be added to plants (or existing ones deleted) contributes to a "separation of the plant from its original environment" (Ruivenkamp, 1989). A better resistance to stress factors makes it possible to shift the geoclimatic limits concerning the growth of specific crops.

TRADE SHIFTS AS THE RESULT OF CHANGES IN FOOD PROCESSING

Important early shifts in international commodity trade will result less from advances in genetic engineering of higher organisms such as plants than from applications of biotechnologies to micro-organisms, a field in which much more experience has been gained. Advances in food processing, especially fractioning plant products into different components and "reassembling" these components into final foodstuffs have led to separation of plants from their specific characteristics. Many crops have become interchangeable. This has greatly increased direct competition between producers of crops that hitherto used to cater to different markets (Ruivenkamp, 1989). The most outstanding example is that of the increasing competition between sugar and starch producers.

The increasing overproduction of maize in North America has stimulated producers' interest in alternative uses of their produce. As a result, research intensified in the 1970s on the enzymatic transformation of starch into high fructose corn syrups (HFCS). The expansion of HFCS production between 1975 and 1985, and the resulting decline in sugar imports by the USA is the largest trade impact that biotechnologies have had hitherto.

Since the early 19th century, it had been known that starch could be transformed into a sweetener. However, advances in the application of immobilized enzyme technology reduced production costs for maize sweeteners to such an extent that a switch from sugar to HFCS became a profitable option - at high domestic price levels for sugar in the USA and Japan. Total replacement of sugar in the USA by HFCS has reached around 6 million tonnes of sugar, with one half of that amount being produced domestically and the other being imported. Sugar imports dropped from 5.3 million tonnes in 1970 to about 2.2 million in 1987. This substitution process has levelled off, because penetration of those areas with favourable technical and economic conditions has reached almost 100%. Large-scale introduction in countries outside the USA and Japan remains to be seen.

While HFCS in its liquid form has been an advantage for industrial users, this very advantage, on the other hand, has prevented the penetration of the consumer market for household use. Only a very small percentage of HFCS reaches the market in crystallized form, because the energy costs of drying made it uncompetitive with refined sugar. Several less expensive procedures have recently been developed, however, which may imply another round of substitution, and could mark the end of sugar imports by the USA.

Another similar example of substitution is the development of improved cocoa butter substitutes (Svarstad, 1988). While such substitutes have been on the market since the beginning of the century, most of them either did not meet the specific desired properties or were too costly for commercial production. With the help of biotechnologies, researchers hope to modify enzymes in such a way that they become fit for the production of cocoa butter substitutes. This would have an impact on the international cocoa trade, if the substitutes were commercially viable and not banned by legislation.

SHIFTS DUE TO THE INDUSTRIAL PRODUCTION OF PLANT COMPONENTS OR SUBSTITUTES

Biotechnologies have increased not only the substitution of one product for another but also the direct competition between agricultural and industrial products. Again, sweeteners provide a good example. Cane and beet sugar lost ground not only to HFCS, but also to industrially produced low-caloric sweeteners, such as aspartam made from two amino-acids coupled by enzymes and now used in many low-caloric soft drinks. Its sweetening power is about 150 to 250 times that of sucrose, depending on the formulation. The world demand for aspartam for more than a decade could therefore be met by a few factories of one company. This is an obvious example of the ongoing process of "dematerialization of production", which implies that the same "use value" can be created with an ever smaller amount of material input. This trend is very much to the disadvantage of raw material exporting countries.

A number of low-volume, high-value compounds which are traded internationally, and which could not (or only at exorbitant cost) be produced by chemical synthesis, can now be obtained through either microbial fermentation or cell culture. This reflects a trend towards the separation of the production of plant components from the land (Ruivenkamp, 1989) and an increasing industrial production of such components. This has a number of advantages: it is less tied to specific seasons but can continue all year long; it can ensure better quality than production which is subject to the vagaries of nature; it normally takes less time on account of the optimization of growth conditions; it is less labour-intensive, because steps like planting, nursing the young plants and harvesting can be avoided; and it allows for easier production than natural processes as is the case of perennial crops, the production of which cannot easily be adapted to market conditions.

The most suitable candidates for industrial production are high-value, low-volume substances such as pharmaceuticals, fragrances, flavours, pigments and insecticides (UNIDO, 1988). A good example is the plan to produce the sweet protein thaumatin with *E. coli,* which could lead to a "substitute for the substitute" thalin, the thaumatin-based sweetener marketed by Tate & Lyle (RAFI, 1987).

SHIFTS AS A RESULT OF THE UNEQUAL DISTRIBUTION OF NEW PRODUCTION PROCESSES

The diffusion of biotechnologies does not take place at the same speed in all countries. Countries that are able to introduce them at a faster pace will consequently be able to increase their own market share, displacing other countries as exporters. Such shifts are now taking place in the field of palm oil and cocoa.

The diffusion of biotechnologies also differs from one crop species to another. Whereas some crops (especially from industrialized countries) have received much attention in international biotechnology research, others have received much less. Where different cash crops compete with each other, breakthroughs with the help of biotechnologies to increase production of one crop species may be achieved at the expense of another. Since the production of different crops is distributed unevenly over different countries, even the equal geographical dissemination of biotechnologies would have uneven effects on the trading position of different countries.

Changing trade patterns

POLITICAL FACTORS

The interaction between technical developments, economic considerations, and political pressure in shaping trade flows is demonstrated by the case of substitution of high fructose corn syrups (HFCS), produced in the United States, for imports of cane sugar. It is estimated that in 1986 HFCS consumption in the USA was around 5.5 million tonnes, and about 1 million tonnes more in the rest of the world, principally

Japan. This concentration of HFCS production is due to the fact that the USA is by far the world's largest producer of maize. This, however, does not explain why the substitution of HFCS for sugar has only taken place in the USA and not in Canada, where almost identical conditions prevail. The reason is that Canada is one of the very few industrialized countries which allows its refineries to buy sugar on the free international market where prices are normally below the world average cost of production. There is therefore no reason for domestic sweetener users to switch from sugar, and virtually all Canadian HFCS production is exported to the USA (International Commission for the Co-ordination of Solidarity among Sugar Workers, ICCSASW, 1987).

Prices for raw sugar in the USA have remained well above world market level since 1981. The average price differential between world and domestic sugar prices has been 353% reaching a maximum of 776% in June 1985 (Maskus, 1989). It was only with the support of the quota system that HFCS production has been profitable in the current period of low world market prices for sugar. US sugar imports fell from 4.54 million tons in 1981 to 2.63 million tons in 1982; import quotas were subsequently further cut down from 2.6 million tons in 1982-1983 to 0.9 million tons in 1986-1987. Three countries account for almost half the total of all quotas (Dominican Republic, Brazil and Philippines), the rest being distributed over 38 other countries.

Internationally, the quota system has been challenged as being incompatible with the rules of the General Agreement on Tariffs and Trade (GATT). The USA agreed to comply with the requirements of the GATT even though it may have significant implications for the US sugar industry (GATT, 1988, 1989). The flexibility to shift back to increased imports of cane sugar, however, is limited, since closures have reduced annual cane-refining capacity in the USA by an estimated 2.5 million tonnes in the past few years (*Financial Times*, 27 September 1988).

The only other major sugar-importing country where HFCS has made important inroads is Japan. Japanese production increased from 84,000 tonnes of HFCS in 1977-1978 to an estimated 650,000 tonnes in 1987-1988 (*International Sugar Journal*, 1989). Japan imports both sugar and maize. Crott (1986) considers that "the inroads of HFCS might have been favoured by a difference of the taxation of sugar and isoglucose in import duties and subcharges". The import of maize to produce HFCS instead of importing sugar has the political advantage of reducing Japan's large balance of payments surplus with the USA. There is no comparable political pressure to reduce Japan's trade surplus with sugar-exporting developing countries. The substitution of HFCS for imported sugar, however, expanded only until 1982 when a surcharge was placed on HFCS. The proceeds were used to help finance the domestic sugar support programme, which resulted in a doubling of domestic beet and cane sugar output to 950,000 tonnes. The surcharge has made sugar more competitive with HFCS, resulting in a recovery of sugar consumption.

The Organisation for Economic Co-operation and Development (OECD) mentions single-cell proteins "from industrial substrates and potential competition with agricultural protein animal feed" as a second example which has present or predictable trade impacts on agriculture (OECD, 1989). Again, political and economic factors are of primordial importance, since the introduction of single-cell proteins and related products is actually "limited more by economic, market and regulatory considerations than by technological constraints" (Litchfield, 1989). Hardly any substitution has taken place up to now, except in the USSR.

EXPECTED IMPACT ON THE TRADE OF INDIVIDUAL COUNTRIES

Some estimates may be presented with regard to the impact of biotechnologies on the trade of individual developing countries. Panchamukhi and Kumar (1988) have presented a rough estimate of the likely annual loss of export earnings as a result of advances in biotechnologies, which could amount to about US$10 billion by 1995 compared to export incomes in 1980.

An obvious negative impact will be the substitution of sugar imports, especially problematic for the Caribbean sugar exporters but much less for Brazil which has some difficulty in producing enough sugar for the fermentation into ethanol and in meeting its export obligations at the same time. Especially hard hit is the Dominican Republic for which income from sugar exports sometimes reached up to 50% of total foreign trade earnings: sugar exports to the USA declined by more than 75% in only six years, from 780,000 tonnes in 1981 to 161,000 tonnes in 1987 (ICCSASW, 1987).

Argentina's meat export declined by about half during the 1980s (IMF, 1988). Biotechnologies will help to increase overproduction in industrialized countries of products that can be used as animal feed. Overproduction in agriculture in general will lead to a more extensive use of soil in large areas, which probably will increase livestock production; it will lead to a better growth of fodder grasses in colder climates. Furthermore, application of biotechnologies to livestock production will reduce the feed/output relation and thus make cattle-raising less expensive in highly industrialized countries. By contrast, an area where exports from some Latin American countries could profit from biotechnologies is the application of mineral leaching to copper production which may be especially suitable for Chile and Peru (Warhurst, 1987).

For the densely populated countries of Asia, biotechnologies may help to achieve food self-sufficiency. This will be at the expense of the large food exporting countries, especially rice exporters like Thailand, Myanmar and Pakistan. The increase of vegetable oil production in India will probably lead to declining palm oil imports from Malaysia.

The Philippines will have to endure the strongest impact on its exports, because two major export commodities are hit at the same time - sugar and coconut products. Coconut oil may be replaced increasingly by cheaper oils and fats on the world market. The decline of income from sugar exports has been already dramatic: while these represented about one quarter of total export income in the 1970s, they declined from a share of 10% to about 1% between 1980 and 1987 (IMF, 1988).

A good example of the chain effects of substitution processes is provided by the impact that the reduction of cassava imports by the European Economic Community from Thailand had on sugar exports: Thailand had to look for new markets for its cassava and increased exports to the Republic of Korea (Berkum, 1988), where starch-based sweeteners already accounted for 18% of total sweetener consumption in the mid-1980s (Crott, 1986).

Some Asian countries will probably be able to increase their share in the total of developing countries' agricultural exports, while applying technological progress rapidly in their production processes. For instance, Malaysia has not only increased palm oil production, but also expanded the large-scale production of cocoa; consequently it became a great competitor to the African producing countries, where cocoa is mostly produced by small farmers. This was not due to biotechnologies, but the resulting large-scale production makes it easier to diffuse any productivity-increasing technology (such as those derived from tissue culture and improvement of varieties).

The victims of the resulting "South-South substitution" will be first of all African countries. Ghana cocoa exports have profited from the special quality of Ghana cocoa, but this advantage may be lost as a result of biotechnological applications, which might be used to "upgrade" less valuable cocoa from elsewhere, so that it matches the quality of cocoa from Ghana.

Applications of biotechnologies to oils and fats will probably also affect the export potential for groundnut oil. Export incomes from groundnut products have shown a considerable decline in Sudan and Mali (IMF, 1988), and may also affect Senegal. The same negative effect may make itself felt in the case of palm-kernel oil exports from countries like Sierra Leone. The production of low-volume, high-value substances (like flavours, fragrances, pigments, insecticides) may have negative effects on countries like Madagascar (cloves, vanilla), Comoros (vanilla), and Kenya (pyrethrum). Sudan has already suffered from the decline of Arabic gum exports.

While the impact of biotechnologies on Latin American and Asian trade will have some negative, but also some positive aspects, African countries will for a long time almost exclusively feel the negative impacts. It will take much more time to use biotechnologies to increase local self-sufficiency in food-importing African countries, because of the many bottlenecks in the dissemination of research results from international research institutions to agricultural extension services and African farmers (Junne, 1987b).

Conclusions

The impact of biotechnologies on trade flows is hard to measure. The main reason is that the impact depends to a very large extent not only on advances in technology, but also on economic factors and on political decisions. Biotechnologies have no direct impact on commodity trade. The influence is always mediated by economic and political variables, such as the strategies of large companies that organize the international division of labour, and political decisions of governments which set the parameters for world trade. Political decisions can promote as well as delay substitution processes. They are also decisive for the regional distribution of the effects of substitution. Biotechnologies will probably strengthen trends that already existed before their introduction, because biotechnological applications have a greater chance of being realized if they serve the major economic and political strategies.

Biotechnologies will first of all affect trade in agricultural products. They will make many importing countries more self-sufficient and increase trade conflicts among overproducing countries. While overall agricultural exports from developing countries will probably stagnate, biotechnologies will help to substitute products from industrialized countries for commodities from developing countries, contributing to a stronger concentration of agricultural production for the world market on fewer developing countries. While newly industrialized countries, given their technological capabilities, will also be able to boost their agricultural production, the least developed countries (especially in Africa and in the Caribbean) will bear the brunt of the adjustment of trade flows.

References

BERKUM, S. VAN. 1988. *Internationale aspecten van het EG-landbouwbeleid. De Relatie met vier ontwikkelingslanden.* The Hague, Landbouw-Economisch Instituut.
CROTT, R. 1986. The Impact of Isoglucose on the International Sugar Market. In: *The Biotechnological Challenge,* Jacobsson, S.; Jamison, A.; Rothman, H. (eds.), pp. 96-123. Cambridge University Press.
Financial Times. 27 September 1988.
GENERAL AGREEMENT ON TARIFFS AND TRADE (GATT). 1988. United States Sugar-trade Policy under Fire. *FOCUS* (GATT Newsletter), 57, pp.1-2.
GENERAL AGREEMENT ON TARIFFS AND TRADE (GATT). 1989. US Accepts Ruling on Sugar Quotas. *FOCUS* (GATT Newsletter), 63, p. 2.
INTERNATIONAL COMMISSION FOR THE CO-ORDINATION OF SOLIDARITY AMONG SUGAR WORKERS (ICCSASW). 1987. *HFCS and Sugar: New Equation in the Sugar Market.* Special publication of Sugar World, Toronto.
INTERNATIONAL MONETARY FUND (IMF). 1988. Balance of Payments Statistics. *Yearbook.* Vol. 39. Washington, D.C.
International Sugar Journal. 1989.
JUNNE, G. 1987a. Automation in the North: Consequences for Developing Countries Exports. In: *A Changing International Division of Labor, International Political Economy Yearbook,* Vol. 2. Caporaso, J.S. (ed.)., pp. 71-90. Boulder, Lynne Rienner Publishers.
——. 1987b. Bottlenecks in the Diffusion of Biotechnology from the Research System into Developing Countries' Agriculture. In: *Proceedings of the 4th European Congress on Biotechnology,* pp. 449-58. Amsterdam, Elsevier.
——. ; KOMEN, J.; TOME, F. 1989. Dematerialization of Production: Impact on Raw Material Exports of Developing Countries. *Third World Quarterly,* 11, 128-42.
LITCHFIELD, J.H. 1989. Single-cell Proteins. In: *A Revolution in Biotechnology,* pp. 71-81. Marx, J.L. (ed.). Cambridge University Press.
MASKUS, K.E. 1989. Large Costs and Small Benefits of the American Sugar Programme. *The World Economy,* Vol. 12, No. 1, pp. 85-104.
ORGANISATION FOR ECONOMIC CO-OPERATION AND DEVELOPMENT (OECD). 1989. *Biotechnology. Economic and wider impacts.* Paris.
PANCHAMUKHI, V.R.; KUMAR, N. 1988. Impact on Commodity Exports.In: *Biotechnology Revolution and the Third World. Challenges and Policy Options,* pp. 207-24. New Delhi, Research and Information System for the Non-Aligned and Other Developing Countries.
RUIVENKAMP, G. 1989. *De invoering van biotechnologie in de agro-industrile produktieketen. De overgang naar een nieuwe arbeidsorganisatie.* Utrecht, Jan van Arkel.
RURAL ADVANCEMENT FUND INTERNATIONAL (RAFI). 1987. *Biotechnology and Natural Sweeteners.* Thaumatin, RAFI Communique. Brandon, Manitoba.
SVARSTAD, H. 1988. *Biotechnology and the International Division of Labour.* Oslo, University of Oslo (Institute of Sociology).
UNITED NATIONS INDUSTRIAL DEVELOPMENT ORGANIZATION (UNIDO). 1988. Growing Compounds from Plants. *Genetic Engineering and Biotechnology Monitor,* 25, pp. 58-60.
WARHURST, A. 1987. New Directions for Policy Research: Biotechnology and Natural Resources, *Development,* 1987, 4, pp. 68-70.

Biotechnology and the public sector

M. Kenney

Introduction

The commercialization of biotechnology-derived products offers potentially important new growth opportunities for less developed countries (see for example, Buttel *et al,* 1985 ; Kenney and Buttel, 1985; Johnston and Sasson, 1986). In every country, the public sector can participate in biotechnology development in a number of ways. Firstly, the government can adopt a more passive role by funding universities to conduct basic or applied research – universities only rarely engage in development activities or marketing of products. Secondly, it can fund the creation of biotechnology companies, or fund existing companies' research. Thirdly, the government can create research institutes for biotechnologies or re-orient existing institutes towards biotechnologies. Finally, the government can influence the development of biotechnologies through its regulating role – for example, decisions regarding safety and environmental issues will have an important impact on the diffusion of biotechnologies.

The government

In the twentieth century, all governments have funded research which in the USA and the United Kingdom has been carried out by the universities, and in Japan and West Germany mostly by government-funded laboratories (Brock, 1989). In most large developing countries such as Brazil, Mexico, India and Nigeria, laboratories and institutes have been created specifically for biotechnology research (Johnston and Sasson, 1986).

In addition to national governments, regional and local government bodies have become actively involved in funding biotechnology activities. This is most noticeable in the USA where nearly every state has underwritten biotechnology research in local universities and firms, on the assumption that the research activities would provide the seeds for economic development based on biotechnologies. These strategies have so far proved unrealistic, as corporate competition has encouraged an agglomeration of firms and biotechnology expertise in Massachusetts, Maryland, California and Seattle. With few exceptions, these local efforts seem doomed to failure, due to inadequate capital, expertise and qualified managers.

In the developing countries the role of local and regional governments is somewhat less in evidence, due to the centralism that affects most of these countries and the concomitant fact that few local jurisdictions have large sources of funds. There are, however, exceptions: the Brazilian State of Sao Paulo, for example, which has allocated significant funds for biotechnologies, notably to the University of Campinas and its Institute of Agronomy; and the Governor of the Mexican State of Guanajuato who has supplied land and funding to support the creation of a plant biotechnology centre (CINVESTAV) in his state. Clearly, non-national jurisdictions can play a role in developing biotechnologies. Whether these local investments will prove successful and valuable for economic development is certainly an open question (Florida and Kenney, 1990). In the developed countries, the most important role of the public sector so far has been to provide funds to universities and research institutes for basic biological research, whereas in less developed countries, universities have emphasized teaching, and the state has generally not funded research on account of the overcrowding of the education system and the overall economic situation.

Another important area of public sector activity is the promulgation of health and safety regulations within national boundaries. The developing countries may be used by developed country firms to circumvent regulations formulated in developed countries. This is a difficult area because stringent regulations stifle the development and use of new products. It may be necessary to develop appropriate international regulations.

The university and biotechnologies

Three different types of university–industry relationship have developed to commercialize biotechnology research, all of which may exist simultaneously at a single university. The first type of linkage is that of a professor who has started a company to commercialize an aspect of research, the second is an institutional affiliation between a single firm and a laboratory or university department, and the third involves the development of biotechnology centres or institutes formed by a number of corporations which often have a component of federal or state involvement. For developing countries, each of these types of relations has both positive and negative features that should be evaluated before being adopted.

PROFESSORS AND ENTREPRENEURIAL COMPANIES

It is difficult to pinpoint when researchers first perceived the possibility that biotechnologies based on the new techniques of genetic engineering and molecular biology could become economically valuable. The watershed event was the formation in 1976 of Genentech Inc., a new start–up company founded by Robert Swanson, a venture capitalist, and Herbert Boyer, a Nobel Laureate molecular biologist at the University of California, San Francisco Medical Centre. They secured the first investment of $ 100,000 from Swanson's former employer, the venture capital firm Kleiner Perkins, and began their research in Herbert Boyer's publicly–supported University of California laboratory. By 1980 when Genentech Inc. went public, both Swanson and Boyer's stakes in the company were worth tens of millions of dollars (Kenney, 1986). The Genentech Inc. experience demonstrated that molecular biology research carried out in the university could yield impressive financial returns to the scientist. In the late 1970s and even more so in the 1980s, university molecular biologists affiliated themselves with or even founded companies. During this period, at least 200 small firms in the USA were started by university professors and venture capitalists to take advantage of university–conducted research. The professors who were closely involved in start–ups received consulting fees, research grants and, most important, equity in the start–up. Obviously, with such enormous capital gains at stake there was great pressure on other professors to use their expertise for personal gain as well (Kenney, 1986). The revolution that occurred in biology has continued to develop new commercial possibilities and thus companies intent on exploiting academic research have continued to be created. At present any top-level molecular biologist in the USA is already connected with some start-up firm.

Although the commercialization of university research has the advantage of moving new developments quickly to the commercial sector, it may also create potential conflicts between professors' academic responsibility and their personal financial situation, and thus risk undermining the scientific enterprise as a whole. The potentially huge capital gains from the commercialization of research results creates enormous pressures for secrecy and can even lead to the falsification of results (Kenney, 1986; Wade, 1984). A similar situation could occur in developing countries where scientists are frequently called upon to assist in the allocation of scarce research funds.

In developing countries, there is far less high technology entrepreneurial activity than in the developed countries, for a number of reasons. Firstly, little venture capital is available and in many, but not all countries, indigenous corporations are unwilling to invest in risky projects. In contrast to the developed countries, the common complaint from scientists in the developing countries is that industry shows little interest in their ideas for new products, preferring to buy a finished process or technology from firms from the developed countries, which means that there is little demand for innovation. In some cases developed country firms actually fund researchers in government-funded research institutions in the developing countries in exchange for access to the products and knowledge generated by that research. Another related problem is that research scientists from developing countries often leave for more lucrative jobs in the developed world. There are some limited signs that university and institute researchers are beginning to form small companies to commercialize their research expertise. In Mexico, for example, a small company specializing in the tissue culturing of flowers has been formed by an entrepreneur with the guidance of university professors. Similar ventures aimed at commercializing new techniques have been started up in Brazil. Whether these little firms will survive and grow to a size that can make an economic difference is certainly open to question. In the USA, most of these small firms tend to be purchased by larger firms seeking the expertise and patents of the smaller firm (Florida and Kenney, 1990).

The exact role that university scientists in developing countries should play in the development of national biotechnologies may be different from the situation in industrialized countries. Large-scale funding of basic research may not be appropriate, given the strain on government budgets. However, individual scientists should most likely be required to receive explicit permission before they can become actively involved with overseas firms. There is a danger that what few assets have been developed through national funding will be transferred to foreign firms.

LARGE LONG-TERM CONTRACTS BETWEEN COMPANIES AND UNIVERSITIES

The rapid development of biotechnologies was a surprise to the large chemical and pharmaceutical companies which had little or no in-house expertise in molecular genetics and thus were unable to assimilate rapidly the commercial potential opened up by these scientific advances. By the late 1970s, though, it had become apparent that many industries would be affected by products developed through molecular genetics. To gain access to the knowledge and skills, it was necessary to gain access to the same pool of professors that the venture capitalists were recruiting to assist in the start-up companies. Contracts between the university and a particular company were therefore established, stipulating that in return for research funding the company would have rights to any resulting products. The corporate aim was to secure access to these new technologies. For the university unit involved the advantage was that research money was available without the necessity of time-consuming preparation of grant applications and the concomitant insecurity regarding funding. For university administrators there was the possibility of charging overhead on the grant and thereby securing financial benefits. In the USA from 1980 to 1983, over $ 140 million was allocated to such long-term contracts, a total which has increased throughout the 1980s (Kenney, 1986).

The largest of the early contracts, for $ 70 million and lasting 10 years, was between Hoechst A.G. and Massachusetts General Hospital (MGH) of the Harvard Medical School, for funding the creation of a molecular biology department at MGH. The privileges that Hoechst A.G. received for this funding were: the right of first refusal for funding any university projects; authorization to prevent department members from consulting with other companies; right of access to all post-doctoral and graduate researchers; the right to have four company researchers in the laboratory at any one time; symposia and research reports for Hoechst A.G.; the right to review all manuscripts 30 days before submission to a journal; and, perhaps most importantly, the rights to exclusive licences on all commercially exploitable discoveries. With this contract

Hoechst A.G. was able to secure access to university researchers (U.S. Congress, 1981).

In contrast to faculty entrepreneurship these long-term contracts formalize relationships between companies and researchers. The advantage, of course, is that the university administration can monitor the contracts. Conversely, however, it provides the private sector with an important opportunity to shape the research agenda, which may in some situations be shifted towards more commercially relevant research. The funding agent has enormous power to influence research directions, either subtly or more bluntly. The long-term relationship permits the dominance by single companies of university laboratories and departments and their research programmes and is accompanied almost invariably by first access and exclusive licenses for the sponsor companies.

THE RESEARCH CENTRE CONCEPT

This system of university-industry contracts may be applicable in developing countries that have large corporations capable of providing such funding and taking advantage of such research. The problem is that these large firms would rather import technology that is already developed. The most salient exceptions to this trend are some of the large firms in North and South-East Asia which have the resources and entrepreneurial managers willing to invest in research and development.

In the USA and Europe in the 1980s, a number of biotechnology centres were started using either national or state funds, and usually including industrial partners. These fall into two contrasting categories – research centres which include, for example, a number of competing pharmaceutical companies, and those which are organized around a group of non-competing companies from different industries (e.g. the Cornell Biotechnology Institute which consists of Eastman Kodak, General Foods and Union Carbide, three companies which are all interested in biotechnologies yet do not compete; Kenney, 1986).

The success of the research centre concept is predicated largely on that of the firms involved to co-operate and on that of university scientists to conduct work of commercial interest to the constituent companies. Nevertheless, the enormous number of centres being developed provide reason for scepticism regarding their success. Often the companies joining these centres see their membership more as a charitable contribution to the university than as a method of acquiring more knowledge. Conversely, the universities and the scientists often see them as little more than another method of obtaining income to fund research.

Evidence gathered thus far indicates that few of the centres have made much progress. Most have been in existence for less than five years, however, and thus it is too early to predict the impacts accurately. The success of many centres seems questionable because of extensive duplication of effort as each state and university attempts to develop its own research centre – all of which may ultimately result in dilution of effort, redirection of resources and opportunity for marginal researchers to garner large amounts of funds. The research centre programme was conceived as a measure to increase the rapidity of information transfer, and yet their performance has been quite irregular.

Developing countries have generally committed their scarce funds to building research centres, rather than adopting the more decentralized approach of funding good projects at a number of institutions. As in the USA, research centres appeal to politicians because they are large projects that generate much attention. These highly visible operations are often handsomely funded initially, but later are provided with insuficient funds to continue operating at the level originally planned. The decision often has to be made whether to continue investing regardless of the research outcomes or to abandon the previous investments to start anew. The cost of this type of research centre development is obvious. In other cases, the research centre is created under the auspices of a renowned scientist or research entrepreneur, which may prove detrimental if the director influences the research agenda though bias, or if the director leaves and the centre loses its vision, political support, or both. Nevertheless, the research centre has the advantage of centralizing scarce human and technical resources, thereby creating the critical mass necessary for success. This large-scale investment is probably necessary to reach the highest levels of science and to induce industry to participate in and commercialize the projects undertaken by the research centre.

PITFALLS IN THE UNIVERSITY–INDUSTRY CONNECTION

It was commonly believed in all the member countries of the Organisation for Economic Co-operation and Development (OECD) that the development of university–industry relationships would yield a more efficient transfer process and increase economic competitiveness. Furthermore, the industry funds could support research personnel and the purchasing of new equipment. Thus, both the university and industry were expected to be winners. This commercialization has had some subtle effects on the way molecular biology is conducted in developed countries. This is because of the tremendous value being placed on biological information. Prior to commercialization, biological materials and information were traded freely. For example, materials which were once exchanged informally are now exchanged only after elaborate disclaimer forms are signed promising that discoveries made with the new materials will not be patented and that the materials will not be exchanged with other researchers. Thus, the information exchange so crucial for academic progress has become truncated and weakened by the injection of commercial motives into traditional academic concerns.

The university's emphasis on patenting discoveries has steadily increased because of the concern among corporate sponsors regarding the appropriability of the benefits of the research that was funded and the possibility that the university will realize income from such discoveries. This increase in patenting injects important new concerns into the research process. Firstly, secrecy is necessary to protect patentability. Secondly, increasing care is taken to establish who should be involved in the research, so as to exclude "non-essential" members. Thirdly, scholarly papers must be written in such a way as to reinforce patent claims. Once again, the injection of these ulterior motives has an impact on the environment and ethos of the university and the very heart of the scholarly enterprise.

The most serious consequences are those related to the collegiality between faculty members. There have been cases in which faculty members have refused or been unwilling to serve on the same committee because of their allegiance to different companies, or to discuss their most recent results at departmental functions for fear that competitors would find commercial value in the advances. In such an atmosphere, distrust and secrecy replace the characteristic scholarly dialogue so necessary for the life of the university (Kenney, 1986). It is not a foregone conclusion that the transference of these types of problems to developing countries would improve the universities' overall contribution to society.

Europe and Japan

The biotechnology industry is international, and as in the Hoechst A.G. / MGH case mentioned earlier, many multinational corporations have invested in American universities. On a lesser scale, American corporations have invested in non-American universities (e.g., investment by Monsanto Co. in protein engineering at Oxford University) (Kenney, 1986). The role of the university in the commercialization of biotechnologies is different not only between Japan and Europe, but also among European countries. It is difficult to make any generalization regarding these very diverse countries except to say that the role of venture capital-financed start-ups have played a far smaller role. Concomitantly, few professors in other countries have been able to capitalize on their specialized knowledge. Thus, the commercialization of university biotechnology expertise has been far more orderly.

In Europe, the university's role in the commercialization of biotechnologies varies cross-nationally. For example, in the United Kingdom there have been both university spinouts and two companies that were formed by the British Government to commercialize certain biotechnologies developed in British universities. With only a few exceptions, however, British industry has shown little interest in research in British universities. The French Government has actively developed some small companies to commercialize biotechnology. In Germany, the universities have proved resistant to entrepreneurial involvement with the corporate sector and few professors have launched start-up firms. However, the large German and Swiss multinational firms have made important investments in biotechnologies and have close relations with the various Max Planck institutes.

Japan has been the most anomalous of all, for until recently professors in the major national universities were forbidden even to consult with industry. The weakness of the venture capital markets also discouraged professors from starting small companies. Thus, the main channels used in the USA to commercialize biotechnologies were not open to Japanese firms, which adopted unique tactics by hiring young molecular biology graduates and dispatching their best researchers to laboratories in Japan and overseas to acquire hands-on capabilities. The Japanese Government has also organized research consortia of companies of university professors which will also assist in technology transfers (Lewis, 1985; Brock, 1989). This method of learning from research by sending personnel for further education was combined with investments in small United States start-up firms. As a result of a concerted effort to catch up with other countries, the Japanese have become major players in the biotechnology arena.

Conclusions

The commercialization of biotechnologies in university-research laboratories led to the creation of new links between the university and industry. This process has had important impacts on the ethos and goals of university research. The involvement of both national and local officials also calls for vigilance. The allure of high technology, large sums of money, and economic development on the basis of research have resulted in new participants guiding research.

The public sector has continued to play a vital role in funding basic biological research. In developing countries there is no doubt that the public sector will provide nearly 100% of all the funding for biotechnologies, which offer many possibilities for allievating the problems facing developing countries. Institutions must be created that can ensure that decisions regarding the research agenda are based on real opportunities. This means the public sector must develop an ability to evaluate projects on the basis of rigorous criteria. Finally, successful deployment of the fruits of research require engineering, manufacturing and distribution. These are not trivial issues, and the effectiveness of the public sector in these fields is open to question.

The importance of the public sector to the development of biotechnologies in developing countries means that the setting of goals and directions in biotechnologies will remain a political process. Systems for evaluating progress and sheltering biotechnologies from political changes will be necessary for any success. In very few nations, except perhaps the Republic of Korea and Taiwan, has such a conscious policy been developed and implemented. Yet this remains the most important single item on the agenda for countries wishing to develop national programmes of biotechnology research for the purposes of international development.

References

BROCK, M. 1989. *Biotechnology in Japan.* London, Routledge.

BUTTEL, F.; KENNEY, M.; KLOPPENBURG, J. 1985. From Green Revolution to Biorevolution. *Economic Development and Cultural Change,* Vol. 34, No. 1, pp. 31–56.

FLORIDA, R.; KENNEY, M. 1990. *The Breakthrough Illusion: Corporate America's Growing Inability to Link Production and Innovation.* New York, Basic Books.

JOHNSTON, A.; SASSON, A. (EDS.). 1986. *New Technologies and Development.* Paris, UNESCO.

KENNEY, M. 1986. *Biotechnology: the University–industrial Complex.* New Haven, Yale University Press.

KENNEY, M.; BUTTEL, F. 1985. Biotechnology: Prospects and Dilemmas for Third World Development. *Development and Change,* Vol. 16, No. 1, pp. 61–92.

LEWIS, H. 1985. Biotechnology in Japan. *Scientific Bulletin* (Department of the Navy, Office of Naval Research Far East), Vol. 10 No. 2.

U.S. CONGRESS, COMMITTEE ON SCIENCE AND TECHNOLOGY. 1981. *Commercialization of Academic Biomedical Research.* Washington, D.C., U.S. Government Printing Office.

WADE, N. 1984. *The Science Business.* New York, Priority Press.

Developing public sector-private enterprise links in biotechnology: experiences in South-East Asia and Australia

P.F. Greenfield

Introduction

The new genetic techniques and bioprocesses which constitute biotechnologies form the basis of new industrial opportunities and the revitalization of some existing industries. Specific commercial developments flowing from these advances have been somewhat slower than was originally anticipated. However, investment in biotechnology research and development world-wide is still growing. Thus between 1982 and 1987, the number of biotechnology patent applications in the USA nearly doubled from 3,116 to 6,153. By January 1988, a backlog of almost 7,000 biotechnology patent applications existed in the U.S. Patent and Trademark Office despite having tripled the number of examiners. The average number of biotechnology patents issued to American companies per annum increased sevenfold between 1981 and 1987. In Australia, biotechnology patent applications increased fourfold from 1982 to 1988 (Biotechnology Committee, Industrial Research and Development Board, Department of Industry, Technology and Commerce = DITAC, Internal Working Document).

Parallels are often drawn between the early development of the semiconductor industry and biotechnologies. Such an analogy is not surprising, in that biotechnologies and semiconductors represent two of the best-known examples of pervasive technologies developed in the second half of the twentieth century. However, several important differences exist between both sets of technologies. Semiconductor technology has given rise to substantial industrial development world-wide, the most celebrated being in California's Silicon Valley. The apparent 'overnight' development of the computer industry to some extent masks many years of fundamental semiconductor research during which time commercial opportunities were relatively restricted. Transistors were first developed in 1947 and integrated circuits in 1959. Chip integration has taken almost 30 years of R & D to produce the Random-Access Memories (RAMs) of today. Biotechnologies are today at the "transistor" stage of semi-conductor development.

Biotechnologies and semiconductor technology development have both been influenced by a number of socio-political factors. Development of the semiconductor industry, especially in the USA, was facilitated to a great extent by the United States Department of Defence. Early products being developed from biotechnology research are mainly health care products and foodstuffs, or else those which involve the release of a genetically modified organism into the environment. Products of this kind are of necessity subject to extensive efficacy and safety testing. Such testing adds substantially to the cost of and time for commercialization of research. Approval of a product by the U.S. Food and Drug Administration (FDA) can for example take up to 10 years and add several million dollars to the development costs of a product. It is only reasonable, therefore, to expect products developed from biotechnology research to enter the market more slowly than was the case for semiconductor technology.

Biotechnology products are, however, beginning to enter the world's food and drug markets at a rapidly increasing rate. Recently the FDA surveyed a group of Japanese companies to determine the number of biotechnology-derived pharmaceutical products being developed: 143 products were identified in various stages of development, the latest product being expected to receive marketing approval in 1998. In Australia, biotechnology-derived products are also beginning to enter the market. Biotech Australia Pty Ltd began marketing its first product and Australia's first recombinant DNA vaccine in 1986. Pac Bio Ltd has produced Australia's first human pharmaceutical by recombinant DNA technology. Bio-care Pty Ltd began marketingNoGall, a genetically modified micro-organism to control crown gall disease in 1989. Numerous other diagnostic products based on monoclonal antibody technology have been developed and are currently marketed by Australian companies both domestically and internationally.

Public sector – Private enterprise links

ROLE OF RESEARCH TRAINING

It is useful to consider a country's strength in biotechnologies along the two dimensions indicated in Table 1. Of significance is the fact that there is no automatic generation in a country of a viable commercial sector based on biotechnologies just because its biological research strengths are high, particularly those in the public sector.

Table 1. A perception of relative commercial and research strengths in biotechnologies.

Research strength in biotechnologies	Commercial strength in biotechnologies: LOW		HIGH
HIGH		USA	JAPAN
		WESTERN EUROPE	
	AUSTRALIA CANADA		
			REPUBLIC OF KOREA
	S.E. ASIA		
LOW			

Australia represents a good example of this phenomenon, having a strong biological research base and enjoying a very high international reputation in a number of key biotechnology research areas. Until the last decade, there was relatively little interaction between the public and private sectors in the commercialization of biotechnologies. In many ways, therefore, Australia faces similar problems to many of the developing countries in establishing viable biotechnology-based industries. It has the potential advantage of a strong research and development infrastructure but, unless this can be focused appropriately and built on commercially, the returns from such an infrastructure will be marginal. Commercial biotechnologies will continue to be driven by advances in the basic biological sciences for the foreseeable future. The immediate products are dominated by human therapeutics because of their potential to return rapidly the large sums of capital spent on research and development and regulatory trials.

The USA represents a good example of a country with high research-and-development performance and a high level of commercial performance. Researchers in the USA learn to apply cutting edge technology to a large number of fundamental biological problems, a number of which are picked up by either the small, modern or larger, traditional biotechnology or pharmaceutical companies. In other words, the mechanism for commercialization already exists. While Japan adopts a more co-ordinated approach, corporate involvement in research from the beginning and in the purchase of international intellectual property ensures a similar result. Within industry, a critical mass of research scientists / engineers and commercial managers / entrepreneurs with access to relevant skills already exists. They provide an important link within existing companies and with the large group of public sector financed researchers in universities and other research centres.

This has not been the case in a country such as Australia nor in most countries in Asia. Apart from Japan, there is no tradition of biological research and development within the private sector, even in those companies which use traditional fermentation or other biological processes. Hence, there is no critical mass of researchers within the private sector with a commercial focus who provide a driving force in their own right and a bridge to the much larger public sector research pool. In addition, their influence partly shapes the culture of the company and its attitude to risk-taking. Faced with such a situation, the role of public sector researchers becomes more difficult. This is because of the need for them to provide the bridge to commercialization, at the very least for an interim period, while the private sector research effort reaches critical mass.

The need for an enhanced role for public sector researchers may have implications for post-graduate and research training programmes. Many South-East Asian countries have programmes enabling some of their top students to undertake post-graduate study at one or more overseas institutions, usually assisted by funds from the home country, loans (Asian Development Bank or World Bank), or aid funds. In biotechnologies, this represents an important mechanism by which new concepts and techniques are transferred to such countries. In addition, the training leads to a build-up of biotechnology researchers, particularly in the public sector.

One difficulty, however, is that the environment in which these concepts and techniques are acquired is generally very different from that in the home country where they are to be applied, particularly as regards the interaction between the commercial sector and the research/ training institutions. Although the techniques learnt are very likely invaluable, the experience in choosing research directions and projects (an essential element of research training) is being gained in an environment which has little in common with that in the home country. The research is therefore carried out in isolation of the needs of the home-based industry. Evidence suggests that the private sector is better suited to choosing research directions and/or projects as it faces the competitive pressures of the market. However, this requires the private sector to have confidence in research and development as an essential component of business, which can be facilitated by collaboration between individual researchers and local companies.

If greater demands are to be placed on public sector researchers to choose research or influence directions appropriate for a particular country, the question may be raised as to whether this can be assisted by modifying post-graduate training programmes. One possible mechanism is for home country institutions to have a greater say in the training of post-graduate students. This can be achieved by means of joint PhD programmes (e.g. organized jointly by a home country university and overseas institution), sandwich programmes (e.g. where the project work might involve an industry-based problem or where the post-graduate programme might involve a structured time in industry, either in the home country or overseas), or by tackling a research problem as part of a larger team (perhaps involving both home country and overseas researchers).

A second important reason for the development of home-based PhD programmes is the creation of a research culture at tertiary institutions. Supervising research requires as much practice as conducting it. The complete absence of post-graduate research students (all studying overseas) both impedes the development of a research culture and limits the opportunities for those who return to learn the skills of research supervision. A number of countries in South-East Asia, including Thailand, Indonesia and Singapore, are now encouraging such programmes in biotechnologies. While it is too early to comment on their actual effectiveness, these appear to be important steps in the right direction.

ROLE OF RESEARCH AND DEVELOPMENT INCENTIVES

In countries such as Australia and those in Asia, industry lacks a tradition of biotechnology research and the critical mass of researchers necessary if a significant economic return from biotechnology research is to be gained. There is a need to provide mechanisms whereby companies can be induced to develop a culture in which research and development, particularly in biotechnologies, is considered an essential activity and component of business, and to utilize effectively those public sector resources in biotechnologies which have been or are being developed.

Within Australia, three such schemes are funded by the Australian Government through the Industrial Research and Development Board of the Department of Industry, Technology and Commerce (DITAC). A fourth scheme, also funded by DITAC, encourages the establishment of international networks of researchers in key fields, including those of biotechnology. These schemes may be summarized as follows.

(a) 150% taxation allowance for approved research and development. This is the most powerful of the incentives because it allows the decision-making on research directions and on the projects to be investigated to remain with the private sector. It aims to change the anti-research-and-development attitude prevalent throughout most Australian upper and middle management, which has been encouraged by 40 years of tariff protection and other market rigidities. Such a scheme must remain in operation for a significant period if it is to have the desired impact. Already pressure is mounting in Australia for the incentive to be reduced or abolished. Within government, the relevant time-scale tends to relate more to election cycles. The increased tax deductibility is of course only of value to those companies which are in a profit-making situation, thereby incurring a tax liability.
The policy is having a growing effect on many industry sectors. This is indicated by official statistics, by the many positions now available to graduating PhD students, and by the increased amount of contract research being awarded by the private sector to public sector research organisations.

(b) Discretionary grants scheme. This provides a grant of up to 50% of the research-and-development costs required to develop a specific product, process or service with fairly immediate market potential. The grants are awarded competitively and are paid on a dollar for dollar basis. They are designed primarily to assist companies which are not in a position to take advantage of the taxation benefits flowing from the 150% allowance for research and development. In particular, this will include small or start-up biotechnology companies. Because such companies are generally cash-deficient, the scheme appears to have led to only limited co-operative activity between the public and private sectors (this is not one of its stated objectives). Its impact on the ability of small companies to maintain a significant research and development effort should not be underestimated, however. The scheme has been widely used in Australia by a large number of local biotechnology companies. The suitability of grants is assessed by referees and a committee, although this may be biased by the need for a committee to pick commercial winners.

(c) Generic grants scheme. This provides up to 90% of the research-and-development costs required to develop a key "enabling" piece of technology in a limited number of generic areas, in this case five, two of which are biotechnologies, and environmental and waste management, respectively. The

stated objectives of the scheme are: to focus research capabilities in tertiary education institutions and government research organizations into agreed priority areas so that meaningful results can be achieved; to encourage the development of enabling technologies, without which significant improvements in critical areas of industrial innovation would be restricted; to encourage collaboration between academics, government and industry in order to obtain the best use of facilities and expertise, to expose each to the capabilities and needs of the other and to develop more market-led research.
In biotechnologies, the priority areas for support were based in part on industry consultation, and are as shown in Table 2.

Table 2. Priority areas for funding in biotechnology generic grants scheme.

Priority areas of research	*Priority areas of application*
Genetic engineering	Plant agriculture
Enzyme application and fermentation technology	Animal production
Cell manipulation and culture	Human pharmaceuticals and medicine
Protein engineering	Food processing
	Alternative uses of agricultural surpluses
	Waste treatment
	Biotechnology support

There is general agreement that this scheme has been one of the major factors in improving the interaction between the public sector biotechnology research and development organizations and Australian industry. There is less agreement on its effectiveness in leading to the commercial development of products, technology, etc. It is clear that the lead times are long.

(d) International science and technology grants. A significant number of grants are now available from DITAC to encourage public sector researchers to maintain or establish contact at an international level. The value of international networking in biotechnologies is now well established. The value of these linkages to the private sector will only occur as a result of the strategies described in (a) to (c) above.

ROLE OF PROFESSIONAL / INDUSTRY ASSOCIATIONS

Professional societies have traditionally been involved with maintaining standards and offering conferences, workshops and journals so that members can remain in contact with developments in their chosen field, at home and abroad. These societies tend either to be discipline-based (e.g. engineering, microbiology) or involved with the technology applied to a particular sector (e.g. water and waste-water). In contrast, trade or industry associations, which have emerged in most countries, have as their objective to promote the commercial interests of their members.

In biotechnologies, scope exists in most countries in Asia and in Australia for the functions of each type of organization to be incorporated into a single association or society. The potential benefits may outweigh the problems of achieving a consensus, which such benefits deriving from the forum for meaningful research/industry interaction which such a society provides. The different needs of the combined membership must of course be recognized, as must the need to associate with the other professional societies which impinge on or utilize biotechnologies.

Such an association has been created in Australia – the Australian Biotechnology Association (ABA). For Australia, with its traditional strengths in biological research but relatively weak history of commercial development, and commercial orientation of biological research and development, such an association was essential. The ABA has been a major factor in the much closer relationships which now exist in Australia between the public sector and private enterprise in biotechnologies. Because of its dual nature, ABA is now recognized by industry and by government as speaking for the whole industry. The current challenges for ABA are to establish better liaison with other more traditional societies and with equivalent organizations being formed in Asia.

Conclusions

For developing countries to appropriate the economic benefits of biotechnology research and development, a strong private sector presence is considered essential. Such a critical mass does not exist in Australia or elsewhere in South-East Asia. Incentives for research and development can have a marked impact on building up this group. In the interim period, a greater responsibility falls on public sector research organizations to interact with industry. This has implications for the training of biotechnology researchers.

Biotechnology and changing comparative advantage: lessons from maize

C. Brenner

Introduction

This assessment of the prospects and comparative advantage for developing countries of introducing biotechnologies in maize production, was inspired by the changing configuration of the economic and institutional environment in which agricultural research in general, and research on agricultural biotechnologies in particular is being carried out, and which is characterized by: the emergence of new private actors (including multinational corporations) and an expanded private sector role in both basic and applied research; budget stringency and / or reduction in government spending and uncertainty regarding the future role to be played by public research and plant-breeding institutions; new patterns of public / private sector collaboration and interaction in research and technology development (Brenner, forthcoming; Goldin, 1990).

Maize was selected as an eminently suitable subject for examining the issues raised by the new environment in which biotechnologies are fostered. One of the world's major crops in terms of volume and value in both production and trade, maize is also an important food and/or feed crop in many developing countries. It has been the object of continuous research effort, reflected in early hybridization, a diversity of food and non-food uses, and continually increasing yields. It may also be one of the first cereal crops to incorporate biotechnological advances. Demand for maize (particularly for livestock feed) which has been expanding in developing countries, often outstripping domestic supply, is forecast to continue increasing in the medium term.

Production and consumption trends in four developing countries

In Brazil, Mexico, Indonesia and Thailand, maize is an important food and / or feed crop (see Table 1).

Table 1. Relative importance of maize in food and agriculture.

	Brazil	Indonesia	Mexico	Thailand
Average annual production (million metric tonnes)	22	6	12	4.2
Annual average production (growth rate %)	2.7	2.4	0.7	4.1
Ranking in terms of area planted	1	2(1)	1	2(1)
Average annual yields (tonnes/hectare)	1.9	1.8	1.7	2.2
Share of food in total maize consumption (%)	14(2)	59	62.2	12(2)
Area planted with improved seed, % of total maize area:				
open-pollinated varieties	7	27	11	84
hybrids	63	3	15	15

(1) after rice.
(2) according to the International Centre for Maize and Wheat Research (CIMMYT).

A wide range of policy measures applied in the four countries directly affect maize production. These include: producer and consumer protection in the form of fixing floor and ceiling prices; seed pricing; fertilizer, pesticide and seed subsidies to producers; the provision of rural credit to poor producers.

Despite a wide diversity of agro-ecological conditions, scale and technological levels of production not only among countries but also among different regions within the same country, certain characteristics of production are common to all four countries. Maize is cultivated mostly in sub-tropical or tropical conditions, under rainfed conditions, on small holdings (e.g. an average of 2.7 hectares in Thailand) and with few chemical inputs.

Overall, recent growth in production has in all countries been due more to increases in yield than extension of the surface under cultivation. Nevertheless, with the exception of Thailand, national average yields are lower than those for all developing countries combined in 1986–88 (2.2 tons per hectare, compared to 6.2 in industrialized countries). National yields can be deceptive, however, as they obscure very wide regional disparities.

In conformity with the trends observed in most developing countries in recent years, total demand for maize has increased in all four countries. This has been due mainly to changing patterns of consumption, where more animal protein and dairy products are consumed and where livestock feed represents a growing share, and human food a declining share, of total maize utilization.

Except in Thailand, which is the only regular exporter of maize grain, maize is produced essentially for the domestic market. Brazil and Thailand both export poultry meat, which accounts for a significant share of domestic maize feed consumption. In Thailand domestic demand for poultry and pig feed has expanded rapidly (around 8% per annum). However, due to government plans to reduce the total area under cultivation to preserve tropical forests, no new land is available. If present demand trends in Thailand persist, it will be necessary either to increase production by 20% (on 75% of the present area cultivated), implying a 50% increase in average yields, or to become an importer. In contrast, since the 1970s, Mexico has become increasingly dependent on imports and is currently the fifth largest importer in the world.

The spread of improved varieties, as indicated in Table 1, varies widely according to country and to region. Little improved seed is used in North–East Brazil, but 70% to 85% of the total surface cultivated is planted with hybrids in the central and southern regions. In Thailand the diffusion of improved seed has been exceptionally rapid since the launching of the open-pollinated Suwan varieties in 1974. Except in Brazil, hybrids still account for only a small share of the total area cultivated. This is in strong contrast to industrialized countries, where only hybrids are cultivated.

Maize research, technology development and diffusion system

Although maize research has been conducted in a more or less organized way in all four countries for at least 50 years, the development of comprehensive research, seeds production and marketing systems is much more recent, particularly in Indonesia and Thailand. Most of the national research and plant breeding effort is conducted by public institutions, although only one country (Brazil) has a national research institution which specializes in maize and sorghum. Common objectives in maize research programmes include: yield enhancement; broad adaptability to semi-tropical, tropical, rainfed conditions; and enhanced pest and disease resistance. To date, although each country has modest biotechnology programmes in other areas, very little public research effort has been devoted to the development of biotechnologies related specifically to maize.

Except in Brazil, where the national firm Agroceres S.A. has spearheaded the research effort, private sector research is limited. In Indonesia and Thailand, it is also very recent. Research conducted by large seeds and agro-chemicals firms, both multinational and national, is often confined to adapting and testing imported seed rather than directed toward developing new varieties based on local germplasm. Most of the small, national seeds firms do not conduct research.

The division of labour, both between private and public sector research activities, and between research on open-pollinated varieties and hybrids is by no means clear-cut. While the public sector concentrates principally on the development of open-pollinated varieties, in all four countries public institutions conduct some research on hybrids. Similarly, the multinational corporations are not involved exclusively in the development of hybrids. In both Brazil and Thailand, some have developed or at least marketed open-pollinated varieties. Since the development of a new hybrid can take 8 to 10 years, open-pollinated varieties are sometimes marketed before the firm's own hybrids appear on the market. This permits the early establishment of a sales network.

In basic maize research, interaction or collaboration between the public and private sectors is either limited or non-existent. In production, multiplication and marketing of seeds, however, interaction varies according to the country, the attitudes regarding the private sector and the maturity of the seeds sector. Regulations such as those regarding research permits, or varietal approval procedures, to which private firms are subject, also differ according to the country.

Contrary to the situation in industrialized countries, even in those developing countries where the private sector performs a significant role in research and in the marketing of new varieties, the government usually plays a preponderant role in the testing, reproduction and quality control of maize seed. In commercial seeds production, the share of the public sector varies widely from one country to another but is predominant in Indonesia, important in Mexico and small in Brazil and Thailand.

The public sector is also involved to some extent in the distribution or marketing of seeds to some extent in all countries. In Indonesia, seeds are marketed through a government agency. In other countries the government is involved indirectly in cases where, for example, rural credit agencies provide credit in kind, in the form of a "technology package" of fertilizer and seeds, usually open-pollinated varieties. Hybrid seed is marketed essentially through private sector firms, national and multinational.

Seeds pricing practices also differ from one country to another. Price information in Brazil is a complex interplay of Agroceres S.A. and Cargill Inc. price setting, market forces and state credit subsidies to agriculture. In Mexico, prices of seed produced by public enterprises are government-controlled, but the private sector is free to set its own prices. In Thailand, prices for open-pollinated varieties are set by the government to serve as a benchmark for pricing hybrid seed. While the range in prices of commercial seed varies according to country, the cost of hybrids is generally at least double and, on occasion, several times the price of open-pollinated varieties.

TECHNOLOGICAL CAPABILITY AND TECHNOLOGY TRANSFER

In Brazil, Indonesia, Mexico and Thailand, the major common constraints in maize research, technology development and diffusion are limited human and financial resources compounded, particularly in Brazil and Mexico, by difficulties in maintaining recent levels of public investment. External funding and various forms of technical assistance, bilateral and multilateral, have been vital to the progress so far achieved both in maize research and in the development of viable seed industries.

It is clear that in all four countries increased productivity could be achieved with existing maize technology through irrigation, mechanization, chemical inputs and improved cultivation practices. Nevertheless, there is a strong need for new varieties better adapted to the diversity of agro-climatic conditions and types of producer. This would necessitate both more systematic exploitation of the indigenous genetic resources available in order to incorporate proven, useful genes in improving plant efficiency, a diversity of location-specific varieties and seed of consistently high quality.

While in the longer term some of the new bioprocesses and maize biotechnologies could certainly facilitate or accelerate the process of producing varieties with sought-after characteristics, it is important to stress two points. Firstly, apart from the scarcity of financial resources, few skills are available in the basic scientific disciplines such as biochemistry, microbiology and molecular biology. Secondly, as most plant biotechnologies will continue to be embodied in seed, they will complement, but not supersede, conventional methods of crop improvement and genetic manipulation, which still require strengthening in all four countries.

In the case of maize, technology transfers to complement local research and breeding efforts are made in the form of germplasm and seeds, of know-how and expertise. The International Centre for Maize and Wheat Research (CIMMYT), which makes improved germplasm available both to national research programmes and institutions and to private firms, has been a key, non-market source of technology transfer for the public maize research programmes of these countries. The limited data available on trade in seeds suggests growing imports of maize seed by some developing countries, including Mexico, now one of the world's largest maize seed importers. Commercial seed imports are important when domestic production is insufficient to meet demand, and can also be used to strengthen or invigorate local genetic resources. However, there are certain critical physical limitations to seed imports which result from the inherent botanical properties of plants. These tend to limit naturally the scope for international trade in seed and to make local production the dominant source of supply for most countries.

The question also arises of the degree to which the maize biotechnologies emerging in industrialized countries are appropriate to the needs and conditions of developing countries. Firstly, they relate essentially to temperate maize and to hybrids. Secondly, they are aimed at quality modification and varieties which conform to very specific quality and nutritional specifications, in response to market conditions where, unlike the situation in most developing countries, specialization and product differentiation are important.

PUBLIC AND PRIVATE SECTOR ROLES

Government intervention is apparent in varying degrees throughout each aspect and phase of the maize research, technology development and delivery system in all four countries. There are, however, strong arguments in favour of public research and of government intervention in other aspects of the technology development and diffusion system, such as the characterization of local genetic resources, the exploration of research directions considered unprofitable or unpromising for the private sector, and the updating of knowledge of genetic improvements in hybrid varieties and, as far as possible, of advances in biotechnologies. There are also particular problem areas either for research or for the diffusion of improved seed where there are no obvious short-term profits to be made, but where the government has a particular responsibility. This is the case with small, marginal producers, often producing for subsistence.

While private sector research is still limited both in volume and scope, the private sector role in seeds production and marketing is expanding. Moreover, in all four countries, multinational seeds and/or agricultural inputs firms are present, although in Indonesia and Thailand their share of the total maize seeds market is small. Private sector involvement in seed markets is increasing despite the disincentives which emerge from the country studies. These include: narrow profit margins for seeds sales; limited diffusion of hybrids compared to open-pollinated varieties; no recognition and enforcement of intellectual property rights; the limited size of the actual or potential seed market in some countries. In addition, regulatory processes related to research authorization, to importing germplasm, to varietal certification, for example, are often cumbersome and lengthy.

One consequence of the indistinct division of labour between the public and the private sectors is that research and development is concentrated in the most productive maize regions which are, at the same time, those best endowed with infrastructure, particularly irrigation. Clearly, there is a need to exploit private/public sector complementarities more effectively and to minimize competition or duplication of effort, in order to make the most of scarce resources, to disperse effort towards geographically disfavoured regions, and to reduce the risks of further aggravating regional imbalances.

INTELLECTUAL PROPERTY RIGHTS

Hybridization is a form of technological innovation which ensures patent-like protection. The improved genetic or other properties embodied in a new hybrid are not transmitted to progeny, with the result that seed replanted a second year will give greatly-reduced yields. Understandably, private firms prefer to develop hybrids rather than open-pollinated varieties in order to recuperate and make a profit on investment in research.

In the highly uniform production conditions of the major industrialized producer countries, the phenomenon of heterosis or hybrid vigour is undoubtedly important in raising yields. However, in developing countries, where most maize is grown under rainfed conditions, with few chemical inputs, farmers' yields of both open-pollinated varieties and hybrids are usually far below those achieved on experimental farms. Evidence from the country studies supports the argument that there is no intrinsic reason for the adoption of hybrids rather than open-pollinated varieties. Results of yield trials reported in the Indonesia and Thailand cases are inconclusive. Where hybrid yields have been demonstrably higher, they have been achieved only with increased application of fertiliser. Often fertilisers are expensive (6 times the price of maize in Thailand) and, in rainfed production conditions, can also imply environmental risk. The cost of producing hybrid seed is also high, and production requires rigorous planting and detassel procedures, carefully supervised seed reproduction and quality control. While progress has been rapid, the production, testing and certification of quality seed are not yet fully established procedures in these countries.

None of the countries has yet adopted a system of protection of intellectual property rights through plant breeders' rights, nor any legislation related to the broader issues of intellectual property rights for plants, animals and micro-organisms, which is an increasingly contentious issue in debates on biotechnologies. Three of the four countries have, however, recently made changes in their patent laws.

Implications for comparative advantage

Major structural changes are already occurring and must be expected to continue in developing country agriculture in the future as a result, on the one hand, of national structural adjustment programmes and policies and, on the other, of the liberalization of international trade in agricultural products. While the likely long-term impact of these joint trends is uncertain, it is reasonable to assume that, as government support for agriculture is reduced, comparative advantage will play a more important role in production and trade than in the past.

It is also reasonable to assume that current trends will reinforce or accelerate some of the long-term changes already under way in agriculture. In industrialized countries, agricultural production has become increasingly input- knowledge-and science-intensive. At the same time a gradual process of "industrialization" of agriculture has occurred, involving, firstly, intensification of the links between agriculture and the non-agricultural sectors of the economy and, secondly, strengthening of the upstream and downstream linkages between agriculture and the seeds, food and food-processing, chemical and pharmaceutical industries. These linkages could be reinforced by new biotechnologies which will increasingly facilitate interchangeability among products and groups of products, and among producers. Mastery of increasingly complex, science-based technologies, including biotechnologies, has become a key instrument in the competitive strategies of firms and governments to maintain or enhance their comparative advantage and shares in markets which are becoming increasingly global.

For developing countries, the structural adjustment and liberalization process implies the privatization of activities and institutions previously under the responsibility of the public sector, as well as diminished government intervention in the economy. It also implies redefinition of private/public responsibilities in research, innovation and the diffusion of technology. This is likely to have both positive and negative effects. Reduction in public expenditure could in some cases lead to greater efficiency in the allocation of scarce resources in the public sector, or in others, to a reduction of resources devoted to crucial long-term research in areas where governments have particular responsibility or interest but no short-term prospects for profitability. Similarly, it may foster complementarity between the public and private sectors and closer integration of the public research effort with what are regarded as strategic market priorities. At the same time, closer public/private sector interaction, in the various forms of contractual arrangements now common in industrialized countries, may imply significant changes in the organization of research, increased private appropriation of research results and reversal of the "public good" criteria which have formerly prevailed in agricultural research. In most developing countries, private sector agricultural research is at present limited both in volume and scope, or even non-existent, and it is unclear what incentives could induce the private sector to assume an expanded role in the development and diffusion of new technology in the short term.

Conclusions

The findings of the maize country study highlight the complexity and magnitude of the task of using new technologies to create genetically improved maize varieties and high-quality seed for a wide diversity of production and agro-ecological conditions. This requires: basic and applied research; efficient seeds multiplication, production and distribution; adequate testing and certification of seed quality.

Progress made and the scientific and technological capability implied in resolving research problems such as that of resistance to downy mildew in Thailand are impressive, particularly with respect to open-pollinated varieties, given that the maize research, technology development and diffusion systems have short-track records, and that ten years or more are needed before any maize breeding programme can give sustained results. However, the problems of ensuring an adequate supply of "appropriate" varieties and of making improved seed available to all types of producers in all major production areas have not been entirely resolved in any of the four countries.

Productivity can still be raised with existing technologies, through perfecting of conventional plant breeding techniques, and through greater efficiency in the seeds industry and seeds distribution. Furthermore, biotechnologies could in future make an important contribution by complementing existing breeding programmes, for example by facilitating the mapping of the genome and accelerating development of varieties designed especially to enhance productivity, lower production costs, modify nutritional content, or improve quality. It would also facilitate, where appropriate, research into alternative uses of maize. The development and diffusion of maize biotechnologies are, however, inhibited by several factors, including the limited scope or even decline in government resources to support the national research, development and diffusion, the scarcity of skills in the basic scientific disciplines on which biotechnologies depend, and the lack of domestic private sector involvement in maize research.

With respect to the international transfer of technology to enhance maize research and breeding programmes, whether in the form of improved germplasm, seed imports or expertise, the International Agricultural Research Centres (IARCs) should continue to play a supportive role. They also, however, face new problems and opportunities stemming from the growing importance of the private sector in the development of biotechnologies. It must therefore be anticipated that access to these technologies will, in future, be more closely linked to the protection of intellectual property rights.

In the long term, biotechnologies may have greater impact on food and agriculture than on other sectors of the economy and will certainly affect the relative importance of different crops and the geographic location of their production; the processing, nutritional content and quality of final food products; and the role of farmers as producers of agricultural raw materials. However, despite the very large sums invested in biotechnology research, except in the field of human and animal health care, few products based on new biotechnologies have yet reached the market. The first major wave of plant biotechnology products is expected between 1992 and the year 2000. It is therefore not clear which biotechnologies will be most profitable to producers of technology or to users, nor which will become most widely diffused. Whether or not biotechnologies will serve to enhance or diminish comparative advantage in agriculture will depend on a number of conditions which include, on the supply side, scientific and technical personnel, equipment and capability necessary to attain the research objectives; and strong capacity in conventional plant breeding, which implies the technical personnel to carry out painstaking but vital long–term field work on experimental stations, as well as capability in process, or bioprocess, engineering.

The conditions likely to stimulate demand for biotechnologies will stem from the impact of economic policy reforms on: exchange and interest rates; price levels, support and variability (for both inputs and outputs); national aggregate supply and crop mix (for export and food crops); levels of investment in agriculture; access to credit and manufactured inputs for different groups of producers; and the technology options available to producers. Lessons from the case of maize suggest that, in the short term, developing countries are unlikely to exploit biotechnologies. This is due, on the supply side, to lack of public resources and scientific and technological capability, as well as to relatively small or undeveloped markets and to low levels of private sector activity in those markets. At the same time, the pressure of demand for biotechnologies is weak, in part because maize has not been (in the countries included in the study) a priority crop in government agricultural policies.

Biotechnologies alone are unlikely to enhance or diminish comparative advantage, or to permit technological "leapfrogging", nor necessarily will they widen the technological gap between developed and developing countries. Rather, they will become another tool to be used in the continuous, cumulative process of technological change in agriculture, which is, in turn, a major component in comparative advantage. This in no way denies the potential of biotechnologies to transform agricultural production and competitiveness in the long term and the consequent need for developing countries to be able to stimulate their capacity to incorporate them when appropriate. It can be argued that, for some countries, the economic and institutional reforms under way, including an enhanced private sector role, will facilitate and accelerate technological change and diffusion in agriculture. For others, however, market forces are still undeveloped and therefore unlikely to provide the motor for technological change in the short term. In those countries, external financial and technical assistance will be required if they are to keep current technological options open and, at the same time, avail themselves of biotechnologies in agriculture in the future.

References

BRENNER, C. *Biotechnology and Developing Country Agriculture: The Case of Maize.* Paris. Organisation for Economic Co-operation and Development (OECD), forthcoming.

GOLDIN, I. 1990. *Comparative Advantage: Theory and Application to Developing Country Agriculture.* Paris, OECD. Technical Paper No. 16 (June 1990).

MATUS GARDEA, J.A.; PUENTE GONZALEZ, A.; LOPEZ PERALTA, C. 1990. *Biotechnology and Developing Country Agriculture: Maize in Mexico.* Paris, OECD. Technical Paper No. 19 (June 1990).

NATAATMADJA, H. *et al.* 1991. *Biotechnology and Developing Country Agriculture: Maize in Indonesia.* Paris, OECD. Technical Paper No. 34 (January 1991).

SETBOONSARNG, S. 1990. *Biotechnology and Developing Country Agriculture: Maize in Thailand.* Paris, OECD. Technical Paper No. 20 (June 1990).

SORJ, B.; WILKINSON, J. 1990. *Biotechnology and Developing Country Agriculture: Maize in Brazil.* Paris, OECD (June 1990).

SUNDQUIST, W.B. 1989. *Emerging Maize Biotechnologies and their Potential Impact.* Paris, OECD. Technical Paper No. 8 (October 1989).

Biotechnology and African economies: long-term policy issues

J.O. Mugabe

Introduction

Agriculture contributes over 50% of the gross national product (GNP) of most African countries and offers employment to over 70% of the continent's population estimated to be 480 million. African agriculture is, however, experiencing major setbacks, as only 30% of its total land area is arable, the rest being arid and semi-arid. The arable land supports about 80% of the population and produces almost all the major cash crops that earn the continent foreign exchange. While the population is increasing at an estimated rate of 3.3% annually, food production in the region lags behind at 1.5%. The level of malnutrition in the region has risen. To feed the increasing population, Africa is forced to import over 20% of cereal foods.

The industrial sector has only marginally contributed to the overall economic progress of the region. In 1986, industry's total output was below 35%. Though the sector is a source of employment and other services, industry has grown at an annual rate of less than 2% in most African countries. Despite this relatively low growth rate, the sector has experienced over-investment. The industrial policies in the region have laid emphasis on protecting local industries from external competition. The quality of industrial products has stayed relatively low, due partly to low levels of technology.

The low growth rates of both the agricultural and industrial sectors have led to serious economic and social problems, such as unemployment, sporadic famines, rising debt and declining capacity to service debt, increasing population but decreasing *per capita* incomes, environmental degradation and declining biological diversity in most African countries. This crisis is the result of interacting internal and external factors: poor macroeconomic policies, instability in political structures, drought, technological stagnation and instability in the world market.

The current structural reforms being advocated by the World Bank and the International Monetary Fund will only deal with the symptoms of the crisis. Dealing with the root causes requires well thought-out, strategic and long-term measures, which could be identified partly through comparison with newly industrialized countries, which have achieved a higher pace of economic development through their ability to harness and utilize technology. The challenge for African countries now is to build up technological, and especially biotechnological, capability. Biotechnologies promise more direct benefits for African economies, such as increased food production, improved health and preservation of the environment. Biotechnologies also, however, pose various threats to African economies. The substitution of tropical products by biotechnological products in the industrialized countries will relocate the production base of most tropical products and erode the chances of African countries to earn foreign exchange. The challenge is to achieve the "appropriate mix", at which African countries will apply and benefit from biotechnologies.

Current status of biotechnology research

Application of biotechnologies in the region is still limited to research activities, most of which are conducted in the national and international research institutions and universities. Although no institution has been established with the sole mandate of conducting biotechnological research, several African countries are planning to set up national biotechnology centres. Biotechnology research in Africa focuses mainly on agriculture, human health and veterinary medicine, the main emphasis being on agricultural biotechnologies for two reasons: the region has a long tradition of agricultural research and agricultural biotechnologies are therefore more easily adopted; research in this area has attracted more donor funding.

AGRICULTURAL BIOTECHNOLOGIES

Tissue culture techniques are the most commonly used. In Kenya, agricultural research is conducted at the Kenya Agricultural Research Centre (KARI), the Faculty of Agriculture of the University of Nairobi, the Faculty of Science of Kenyatta University and at various international research organizations. At the Department of Crop Sciences, University of Nairobi, meristem culture and embryogenesis are applied on a number of crops, especially coffee and horticultural crops, to develop pest-resistant plantlets. Tissue culture on maize is conducted at Kenyatta University. Certified potato tubers using tissue culture techniques are being developed by the KARI National Plant Quarantine Station at Muguga, and the National Potato Research Station at Tigoni uses tissue culture for clonal propagation of potato varieties. The elite clonal material is then passed on to farmers. The International Potato Center (CIP) provides technical support for a collaborative potato research programme with the KARI.

Another area of biotechnology research in the region is nitrogen fixation, carried out under the auspices of the Microbial Resources Centres (MIRCENs) network, sponsored by UNESCO. Research in this area has already resulted in the commercial production of *Rhizobium* inoculants at the University of Nairobi's Department of Soil Science. The main objective of this project is to promote the use of biofertilizers through field applications and to reduce environmental pollution resulting from the use of inorganic fertilizers. However, it lacks an effective extension system.

In Ethiopia, which has greater genetic diversity than any other area of sub-Saharan Africa, tissue culture is used for the conservation of germplasm at the gene bank at Addis Ababa. Similar programmes are conducted at the gene bank of Kenya, run by KARI at Muguga.

Environmental conservation is emerging as another area of biotechnology research. Forestry biotechnology is an important activity at the Kenya Forestry Research Institute (KEFRI) where tissue culture is used to multiply indigenous tree plantlets and to generate disease-resistant varieties to be used in reforestation programmes.

Zimbabwe has the most advanced biotechnology research programme in the region, as its infrastructure for research is better developed than that of the majority of sub-Saharan countries. Zimbabwe also co-ordinates the food security programmes of the Southern African Development Co-ordination Conference. At the Department of Crop Sciences, University of Zimbabwe, tissue culture is used to develop disease-free varieties of tomatoes, potatoes and coffee. Using the

Staritsky leaf disc technique, elite coffee bushes have been cloned. The Tobacco Research Institute of Zimbabwe conducts research to incorporate resistance to tobacco mosaic virus into tobacco using pollen culture and somaclonal variation.

At the University of Burundi, researchers are working on *in-vitro* propagation of sorghum and maize. The university has also established programmes for the multiplication of yam, potato and cassava. In Nigeria, the main areas of research concentration are crop production and storage. Research efforts have led to the production of high-yielding and disease-resistant varieties of cassava and yam; these varieties are disseminated to farmers with relevant information for production.

VETERINARY BIOTECHNOLOGIES

Veterinary biotechnology research is mainly conducted in the International Laboratory for Research on Animal Diseases (ILRAD) in Nairobi, Kenya, on East Coast Fever (ECF) and typanosomiasis; it has developed diagnostic tools aimed at producing vaccines against these two diseases. The institution recently developed species-specific monoclonal antibodies against three African typanosome species infecting cattle. Further research is carried out on genetic engineering to develop molecular hybridization reagents for epidemiological studies of *T.congolense* and [031] *T.vivax,* with the aim of identifying various genes that code for resistance and transferring them to the cattle germline. ILRAD has designed a restriction map of the genome *T.parva parva* which is organized in four chromosomes. All these research activities are aimed at controlling typanosomiasis, which accounts for considerable loss of cattle every year. At the International Livestock Centre for Africa (ILCA, Addis Ababa), techniques such as enzyme-linked immuno-sorbent assay (ELISA) are being used in disease diagnostics.

In Nigeria, the major research activities include production of bacterial and viral vaccines for livestock protection against a wide range of tropical diseases. Most of the veterinary biotechnology research is conducted by multinational corporations operating in Nigeria.

HUMAN HEALTH BIOTECHNOLOGIES

Various biotechnologies are being used as diagnostic tools for tropical diseases. One of the most recent achievements is the development by the Kenya Medical Research Institute (KEMRI), through its research on alpha-interferon, of a drug – Kemron – which is reported to clear AIDS symptoms effectively within a few days after administration. It is administered at low doses of 100 units and was developed through collaborative research between KEMRI, an American cell culture firm (Amarillo), and a number of Japanese researchers. The development of Kemron and the controversies that arose thereafter, notably those regarding patents, clearly indicate how the absence of effective patents in the region will continue to hinder technological innovations.

At the University of Zimbabwe, a *Salmonella*-DNA specific probe has been developed and is used to test for salmonella in veterinary services and foodstuffs. Research is also being conducted to develop diagnostic tools for hepatitis B virus.

BIO-INDUSTRY

Bio-industrial activities are still limited, due perhaps to the fact that the industrial sectors of African countries are dependent on foreign expertise. Moreover, a large percentage of the industries are owned by foreign firms which draw most of the research needed for their manufacturing activities from their mother firms abroad.

In Nigeria, activities relate to the development of palm wine, champagne from kola, and beer from sorghum and millet. Biotechnologies could also be used to process cassava industrially into animal feeds, starch and alcohol products, to improve the quality and volume of production in the baking industry, and to enhance the fermentation processes.

Opportunities and challenges

Plant biotechnologies could help reclaim arid lands through the development of drought-resistant crop varieties, as well as reforestation programmes, thereby reducing environment degradation. Bio-industry can help in upgrading the quality of local products. However, if these benefits are to be realized, African countries will need to harness their resources and build endogenous technological capability.

CONSTRAINTS

Effective biotechnology programmes will not be realized until there is sufficient political will to enhance research and the application of biotechnologies.

Africa has a shortage of qualified personnel in science and technology; on average, each university produces only five graduates per year with Ph.Ds in science – and technology-related subjects. Training is constrained by lack of funding, and shortage of staff and equipment. Only the University of Zimbabwe has introduced a course on biotechnology at the M.Sc.level. It is estimated that Africa has less than 700 personnel engaged in biotechnology-related research, an average of 10–12 persons per country, who, moreover, are scattered throughout various institutions, with little possibility of co-ordinating their research activities. The following table shows that less than 40% of scientists in ten countries have postgraduate training in agricultural research.

Staff qualifications in national agricultural research institutes

Country	*B.Sc*	*M.Sc*	*Ph.D*	*Total*
Botswana	7	11	2	20
Ethiopia	129	59	14	202
Kenya	246	205	25	476
Lesotho	6	11	0	17
Rwanda	16	6	2	17
Somalia	33	7	0	40
Tanzania	147	134	27	308
Uganda	106	78	16	200
Zambia	51	33	4	88
Zimbabwe	109	54	32	195
Total	850	598	122	1,570
Percentage	54.1	38.1	7.8	100

Source: Mkiibi, J.K.; Omari, I. (eds.). 1989. *Training for agricultural research and training in Eastern and Southern Africa.* Nairobi, Initiatives Ltd.

It should also be noted that where qualified personnel exist, they are not effectively utilized. For instance, in Tanzania, only 13% of staff in the research institutes carries out research work, the rest being used in administration and other support activities.

The quality and availability of infrastructures relevant to biotechnology research is relatively low. The high cost of equipment places it beyond the means of most research institutions. For instance, the cost of a simple autoclave is over 200,000 KShs, about 10% of the total budget of a research department at the University of Nairobi. Moreover, lack of institutional flexibility and interdisciplinary collaboration makes the management of biotechnology programmes difficult.

The government budget for research has remained low in most African countries. In Uganda, less than 5%, and in Kenya and Tanzania, less than 20% of the total government budget is spent on research. Funding for most research activities is drawn from financial and donor institutions, such as the United Nations Development Programme, the World Bank, the International Development Research Centre (IDRC, Ottawa) and a number of non-governmental organizations located in the region. Funding in most instances is limited and unstable since biotechnology programmes are usually dependent on the initiative of a few people; institutional instability and the lack of consistent measures in favour of biotechnologies make donors hesitant to provide substantial and long-term funding for biotechnology programmes.

A further constraint is the lack of information on biotechnologies. There is no institution that has developed capability in information collection, retrieval and dissemination. So far, the region relies on institutions located in the industrialized countries for the acquisition of specialized information. Policies on benefits and risks of biotechnologies can only be formulated if information on those impacts is readily available.

CHALLENGES

Biotechnological progress in the industrialized countries is likely to weaken African economies if strategic measures are not taken by African countries to protect themselves from some of the negative consequences. The main threat lies in the anticipated displacement of some of the cash crops grown in the region or the substitution of tropical products, which earn foreign exchange. Coffee exports from Uganda, Tanzania and Kenya may, for example, be affected by the application of current research in South-East Asia to increase coffee production. Pyrethrum, a major export product of Kenya and Tanzania, may be similarly affected by current research at the University of Minnesota, USA, to produce pyrethin by cell suspensions instead of importing it at the present price of $ 200 per kilogramme. Another tropical product targeted for substitution is vanilla, which represents over 25% of the export products from Madagascar and 34% of those from Comoros. On average, these two countries earn $ 70 per kilogramme of high quality vanilla. The International Plant Technology Institute (USA) and Firmenich (Switzerland) are working on the production of vanilla by cell and tissue culture.

The relocation of production bases will cause unemployment in the producer countries and destabilize their economies. Options for African countries using biotechnologies to improve and sustain the quality of the export products do exist, and such measures may enable them to adapt to changes in the international trade patterns.

TECHNOLOGY TRANSFER

Africa imports almost all of the technologies it requires in the agricultural and industrial sectors, and has not yet built an indigenous technological capability. One important aspect of technology transfer is the bargaining power of the buyer and the prevailing conditions for importing the technology in question. Most African countries have low technological capability and limited purchasing power. The cost of purchasing technologies is high, and most African countries have no adequate infrastructure to enable effective use to be made of the imported technologies.

Long-term policy options

POLICY RESEARCH

There is need to conduct policy research to enable African countries to identify the positive and negative impacts of biotechnologies on their economies. So far, very few impact studies have been conducted.

Policy research on biotechnologies calls for an interdisciplinary approach covering biology, bio–engineering, sociology and economics. Furthermore, African countries should monitor biotechnological developments in the industrialized countries and identify those biotechnologies that are socially acceptable, and economically and technically feasible.

INFORMATION MANAGEMENT

One way of dealing with the problem of lack of information is to conduct a thorough literature search with a view to identifying the existing information, which should then be stored in the institutes having the capability to manage the information and to make it easily accessible to biotechnology researchers. The other option would be to set up a biotechnology information centre in the region, by establishing a computer link with institutions in the industrialized countries. This depends on the willingness of the institutions in the North to share their biotechnological information with those in the South. This is an aspect that UNESCO and other international organizations might possibly introduce into their programmes for Africa.

TRAINING, TECHNICAL SUPPORT AND INSTITUTIONAL CHANGE

Ad hoc training programmes for researchers can be implemented in the region through collaboration with research and training institutions in the North. Researchers from African institutions could go to Europe, Japan or the USA for short periods and return home to apply the knowledge they have acquired; they could return for re-training so as to keep abreast with technological developments.

The other approach is to develop training capabilities within the existing local institutions. This could be done first by including biotechnology subjects within current degree courses. It might be appropriate to have technical personnel from institutions in the industrialized countries work together with the local staff. These countries may help to develop technological capability in Africa by offering training for African researchers at reasonable cost, as well as technical assistance in biotechnology research.

There is need for an inventory of the existing institutions engaged in biotechnology research, which will help to identify the weakness and strength of particular institutions in the region, with a view to carrying out reforms. Furthermore, new forms of partnership should be developed between local institutions and also with institutions in the industrialized countries, with a view to facilitating the acquisition and sharing of biotechnology information.

International support is also required: funding and technical expertise for current research programmes; co-ordination of training programmes on biotechnologies in African universities and overseas; technology transfer and development of indigenous capability to acquire and apply specialized information.

Conclusion

Current efforts made in biotechnology in different countries should be supported by African governments and the international community. Policy studies are required to identify priority areas of research and to adopt measures that will enable African countries to reap the benefits of biotechnologies and at the same time to protect themselves from some adverse effects. African countries should not adopt wait-and-see attitudes, but should take the initiative themselves.

Will biotechnologies be a threat or an opportunity for the south?

A report on the current status and future targets for biotechnology-aided development in Africa, in particular Zimbabwe

I. Robertson

Introduction

The Industrial Revolution set Europe and then America on the road to economic dominance. Then the communications microchip revolution lifted Japan and the Pacific Rim into some prosperity, riding initially on Western know-how. Then the "green revolution", based on much-improved varieties of rice and wheat, helped Asia out of poverty into self-sufficiency; yet Africa and Latin America lag behind; their staple crops are of course different – cassava, sweet potato, beans, groundnuts, maize and millets – but the "biotechnological revolution" now opens up new horizons. Unfortunately, few African countries have the skills, infrastructures, administrative know-how, or the confident capital to use even hand-me-down technology. Yet, the benefits of biotechnologies can be made readily available and "indigenized" with relatively few scientists and a modicum of equipment. To make progress, African countries need to earn the confidence of the decision-makers and to develop structures free from undue bureaucratic delays. Biotechnologies develop and change fast: if African countries are slow to respond, they will not be competitive and will continue to lag behind. The question remains as to whether biotechnologies can help to combat the poverty of these countries and whether, being powerful tools, they are going to be a threat to their economies or the vehicle for economic liberation.

Africa has 400 biotechnologists and technicians, Latin America

2,000, Japan, 4000, while the USSR has 12,000 and the USA 23,000. Africa's 400 are scattered over 60 countries (Edroma, 1990). Clearly there is need for more, and greater concentration into critical masses.

Plant tissue culture can be found in about a dozen African countries, commercial usefulness being limited to perhaps five of those. Another handful have a plant for producing nitrogen-fixing bacterial inoculants. Very few produce even their own yeast cultures for brewing, or spawn for mushroom cultivation. Zimbabwe has just begun to grow her own hops aided by tissue culture propagation (Sakina, 1988). Kenya has aroused interest and controversy with an AIDS therapy (Koech *et al.*, 1990; Obel, 1990), but it is too early for the results to be evaluated. A number of Africans have done fine work in DNA recombination abroad, but lack the facilities to continue the work at home. This kind of work can be found only in Egypt, Kenya, Zimbabwe and South Africa; it is emerging in Cameroon, Nigeria and Morocco.

Threats to domestic production and international trade

Biotechnologies may constitute a threat to domestic production in developing countries by increasing the penetration of their export markets. This is already happening in the production of sugar, and could happen for coffee (Rural Advancement Fund International, 1989) and vanilla. It could happen soon in the production of low-caffeine, flavour-improved coffee, test-tube cotton fibres (Balinger, 1989), and high-grade cocoa.

The domestic market in these countries may also be affected by the development of pest-resistant and virus-resistant varieties of maize, cotton, rice, potatoes, cassava and sweet potato. Applied biotechnologies may thus reduce inputs in the industrialized countries, which can then produce the crop more cheaply and export it to developing countries, which would in turn suffer from increased unemployment and decreased foreign currency. Another risk is that African countries may lose control of their own Seed Coop that produces hybrid maize and other quality seed. If rival companies produce the above new varieties, then the home-based seed business will be destroyed or taken over.

The new opportunities created by biotechnologies will, of course, benefit the industrialized countries first. One disturbing novelty is that the micropropagation system has begun to be automated in the USA and Japan. This is an advantage for the industrialized countries where labour is expensive, but not for the Third World, with its large, available labour pool. The terms of trade are currently deteriorating. With biotechnologies in the North and no biotechnologies in the South, they can only become disastrous. The only possible response is for the South to develop its own biotechnology capacity.

The world-wide graph of fertilizer sales shows an exponential rise from the early 1940s onwards. The rate of increase slowed from the late 1960s, but is still rising steadily. Herbicide, insecticide and fungicide sales took off dramatically from around 1970, and all three are still climbing exponentially (*New Scientist,* 1989). The combined volume of those three biocides is around 600 million tonnes per year, which generates substantial profits. Seeds of varieties that react better to one or another biocide could tip the balance with farmers'choices and shift the market share to one or another biocide product. There is the danger that eventually farmers will be required to buy the seed along with the associated biocide. This increasing use of fertilizers and biocides could be disastrous for environmental reasons, affecting mineral balance, soil crumb structure, ion binding capacity, organic turnover and micro-organism viability.

How can the developing countries respond? A successful *Bacillus thuringiensis* (Bt) gene in corn would save the American farm $ 240 million a year. When that technology trickles down to African countries, they may find that industrialized countries can grow it cheaper (really cheaper, not subsidized cheaper) than they can. Then they may have to import it despite being able to grow it themselves – a chaotic situation. A successful Bt gene in rice, saving about 1% of the crop, would save $ 300 million every year for India alone (Bennet, 1991). In maize, in Africa, at least $ 100 million could be saved because stem-bore destroys at least 10% of the crop. Which government has realized that countries in the South could dramatically reduce costs by investing in modern biotechnologies? Who is ready to put up the first 1% of the likely saving? Who has calculated the cost of not doing so?

Opportunities for growth

The only sane hope for countries in the South is to compete by building the capacity to do the job themselves. As with self-sufficiency in food, these countries must aim at self-sufficiency in technology. It must be borne in mind that effective biotechnologies depend on only a small number of individuals doing something of enormous significance, using cheap and powerful tools. The cost/benefit ratios could be staggeringly good. The critical mass of the required scientists can be very small. The infrastructure for delivery depends on politics, culture and honest administration, but the medium of delivery is a bag of seed and a bundle of stem pieces, and those take very little sophistication for dissemination. They will also create their own demand.

PRACTICAL PROPOSALS

There must be at least one tissue culture laboratory in every developing country, for import and export of new clean or transformed cultivars; for clean-up of old, diseased, but potentially good traditional cultivars; and internal large-scale propagation of good material.

There must be regional centres for DNA recombination. With certain proven, on-stream technologies, the potential is there now to double the yield of every staple crop with, in addition, about 20% reduction in input costs. That would provide self-sufficiency in food, fibre and fuel biomass. The infrastructure for delivering the seed to the needy does not yet exist, of course, but it could be developed. An effort must be made to inspire venture capital with confidence in Third World expertise.

VENTURE TARGETS

A list of potential investments would be as follows.

One million dollars would suffice to deliver all the insulin Africa needs, grown either in tobacco leaves or sweet potato tubers.

A second million dollars would establish ten tissue culture facilities in countries where none exist so far. The scientists could be trained in Zimbabwe, Cameroon or Kenya. Malawi, Botswana, Mozambique, Angola, Zambia and Zaire could all benefit from these laboratories passing on good new cultivars to their farmers.

A third million could be invested in modifying the amino-acid balance and the poor protein content of cassava and sweet potato. Those are the crops which the poor of Africa will be eating in the future. Terms of trade can also help, as one ECU or dollar goes about five times further in Zimbabwe than in the North, in terms of work done and salaries paid.

In Zimbabwe, a good hop variety for the brewing industry has been imported through tissue culture. The national brewing company has learnt how to grow the hops so as to control their photoperiod, which saves about a million dollars in hard currency each year. A fourth million dollars could help to transfer that proven technology to other countries and adapt it to local temperatures and photoperiods.

A fifth million could be invested in digging out an ant-resistance gene from local African plants; the methodology is there and imitation is cheaper than innovation.

A sixth million could go to developing Kenya's Kemron, which shows some interesting preliminary effects on AIDS patients. If this proves effective, it would be worth keeping the therapy in the hands of those who first thought of it and tried it out on needy human beings.

A seventh million could go to producing drought-tolerant sorghum, millet and maize. Salt-tolerance has been achieved through somaclonal variation and there is a theoretical basis for the idea that such variants could be found that are drought-tolerant and able to survive on residual moisture. If planted after erratic rains have fallen, they could reduce considerably the risk of living in semi-arid regions, which include half of Africa.

Conclusions

The beauty of the biotechnological revolution is that ideas are free, transnational, and very contagious. Equipment can be added for a million dollars. If it is obvious that poverty-threatened nations would be able to save tens of millions of dollars a year by using such processes, who would deny them the equipment? The only factor missing is the critical mass of scientists. A busload of dedicated, committed, funded scientists, plus a container-load of equipment could soon cope with the pressing problems, the crushing poverty and hunger, and the intellectual loneliness of the Third World.

The debate regarding patents and Plant Breeders' Rights continues, although plant genes should be free and the benefits shared by all. There are, of course, dozens of products of unpatented genes out there in the jungles and grasslands of the Third World that could if necessary be sought out, isolated, worked on, and incorporated into our crops: the cow-pea trypsin inhibitor (CPTI) gene, legume lectins, chitinases of several types, inhibitors of trypsin and chymotrypsin could all be used. There is truly enough genetic potential in the world for all our needs, and our main task now is to collaborate in sharing it out.

References

BALINGER, A. 1989. Is there a Future for Super-Cotton? *International Agricultural Digest*, September 1989, January 1989.

BENNET, J. 1990. United Nations Conference on Science and Technology for Development / African Biosciences Network; Conference on biotechnology for semi-arid lands (Dakar, 8–11 October 1990).

Edroma, E. 1990. Introductory Address, UNCSTD/ABN Conference, see above.
Koech, D.K.; Obel, A.O.; Minowada, J.; Hutchinson, V.A.; Cummins, J.M. 1990. Low-Dose Oral Alpha-Interferon Therapy for Patients Seropositive for Human Immunodeficiency Virus Type-1 (HIV-1). *Mol. Biother*, 2, pp. 91–6.
New Scientists. 15 July 1989, p. 15.
Obel, A.O. 1990. Keynote address. *Proceedings of the 2nd Zimbabwean Symposium for Science and Technology.* (Harare, October 1990).
Rural Advancement Fund International (RAFI). 1989. *Communique, Coffee and Biotechnology.* Austria, Vienna, July 1989.
Sakina, E.K. 1988. Personal Communication on Micropropagation of Hops for Delta Corporation.

Incidences socio-économiques du développement des biotechnologies dans les domaines agricole et alimentaire en Tunisie et dans la région sud-méditerranéenne

M. Ben Said

Introduction

Le Maghreb et le Machrek, groupe des huit pays du Sud de la Méditerranée (Maroc, Algérie, Tunisie, Libye, Égypte, Syrie, Liban, Jordanie), demeurent confinés dans une spécialisation primaire, important des biens manufacturés et des denrées alimentaires, exportant des produits miniers et énergétiques. Traditionnellement exportateurs de produits agricoles, ces pays sont progressivement devenus des importateurs, enregistrant ainsi un solde négatif important de leurs échanges agricoles notamment avec la Communauté économique européenne (CEE).

Les importations de denrées alimentaires contribuent pour plus de la moitié des besoins en moyenne (trois quarts en Algérie et en Jordanie, plus du tiers en Égypte). Pour l'ensemble de ces pays, la facture se monte à plus de $ 10 milliards en 1984, dont plus de la moitié concerne l'Égypte ($ 3,7 milliards) et l'Algérie ($ 2 milliards). Cette évolution, amorcée au début des années 1970, place les pays dans une situation de dépendance extérieure structurelle à long terme.

Le poids croissant de la facture alimentaire a conduit les gouvernements des différents pays méditerranéens à définir une nouvelle politique agricole pour tenter de réduire les coûts des importations. A cet effet, il fallait augmenter les prix à la production afin de promouvoir la production intérieure. Par ailleurs, la pression des populations contraint les autorités à maintenir l'approvisionnement du marché à un niveau supportable. Ce double soutien à la production et à la consommation s'est traduit par des coûts qui hypothèquent l'effort d'investissement et la création d'emplois. En même temps, la suppression progressive des subventions aux intrants agricoles a entraîné une augmentation des coûts de production. Ces mesures relèvent, pour l'essentiel, de la sphère de l'échange, c'est-à-dire la manipulation des prix, alors que les véritables problèmes de l'agriculture sont relatifs à la capacité de production et à la maîtrise des technologies.

Un autre aspect de la politique agricole concerne les exportations agricoles (Maroc, Tunisie) vers la CEE, qui sont gênées par l'intégration des pays de l'Europe méridionale (Espagne, Portugal) à la Communauté. Les pays méditerranéens de la CEE accentuent leur spécialisation dans les cultures méditerranéennes (huile d'olive, agrumes, horticulture) aux dépens des cultures vivrières pour lesquelles les pays septentrionaux sont plus avantagés.

Biotechnologies et production agro-alimentaire

En Tunisie, le Centre de biotechnologie de Sfax, créé en 1983, n'a commencé effectivement ses travaux qu'en 1988. Ces derniers portent, en premier lieu, sur l'hydrolyse de l'amidon du gruau pour obtenir un sirop d'isoglucose ou fructose, de façon à réduire le déficit sucrier de la Tunisie, qui ne couvre que 5 % de ses besoins. L'hydrolyse et l'isomérisation sont effectuées à l'aide d'enzymes (amylase, amyloglucosidase et glucose-isomérase), extraites de *Streptomyces violacoeniger.* En deuxième lieu, la digestion anaérobie par voie microbienne des margines qui sont des déchets de l'extraction de l'huile d'olive, a permis de réduire de 90 % environ la demande chimique en oxygène (DCO). En troisième lieu, la mise au point d'un procédé optimal de production de lysine pour les rations alimentaires des animaux d'élevage, a consisté à rendre optimales les conditions de production de lysine sur hydrolysat de gruau de blé ; à construire des vecteurs-navettes autorisant l'expression de gènes chez les corynébactéries d'intérêt industriel ; à améliorer, par le génie génétique, les souches productrices de lysine. En quatrième lieu, des travaux sont menés sur la biodégradation des matières ligno-cellulosiques et pectiques, en vue de valoriser les déchets ligno-cellulosiques (dégradation de la cellulose en glucose ; utilisation des cellulases dans les industries textiles, la fabrication du papier et les lessives ; amélioration de l'extraction de l'huile et de la digestibilité des aliments du bétail).

La propagation du palmier dattier (variété Deglet Nour) à l'aide des cultures de tissus vise enfin à étendre les plantations et à multiplier les variétés résistantes au bayoud, maladie provoquée par le champignon *Fusarium oxysporum* f. *albedinis.*

Les travaux menés depuis 1988 à l'Institut national agronomique de Tunisie (INAT) portent sur l'utilisation des mycotoxines pour la sélection *in vitro* de variétés de pois chiche résistantes à l'anthracnose causée par l' *Ascochyta scabiei,* en collaboration avec l'Université de Londres et le Centre de biologie et biotechnologie du Pakistan (Lahore) ; sur l'utilisation de la méthode RFLP (Restriction Fragment Length Polymorphismé ou polymorphisme de la longueur des fragments de restriction de l'ADN, pour repérer les gènes contrôlant des caractères agronomiques et pour opérer la sélection variétale de façon plus rapide et plus précise ; sur l'inoculation simultanée d' *Agrobacterium rhizogenes* et de *Rhizobium* chez le lupin et la fève, en vue d'augmenter le chevelu des racines et le nombre des nodosités.

Conclusions

Les pays en développement du bassin méditerranéen pourraient craindre que les biotechnologies élargissent davantage le fossé avec les pays industrialisés. Un nouveau style de coopération et d'autres modalités de transfert de ces technologies pourraient, en revanche, conduire à plus d'optimisme. Un programme intensif de formation et de recherche-développement permettrait en particulier de consolider les structures de la recherche scientifique et technique et les moyens d'intervention des pays, et de répondre aux besoins du développement agricole et agro-alimentaire. D'autre part, en considérant l'évolution des rapports entre la CEE et les pays tiers méditerranéens, et en tenant compte de l'accroissement des effectifs des populations, la CEE semble prendre conscience de la nécessité de mettre en oeuvre une stratégie globale de co-développement avec ces pays, qui irait au-delà des aides et des échanges traditionnels.

Possible impacts of biotechnology on Venezuela's agro-industry

A. Martel

Venezuela's agro-industry

Food products represent two-thirds of the total value of goods sold in Venezuela. In the manufacturing of goods by private industry approximately 40% of the aggregate value comes from the agricultural food-producing system, which constitutes 20% of the gross national product (GNP) and is surpassed only by the oil industry. Furthermore, the agricultural food-producing system is of vital importance for the stimulation of the economy, as it creates stable employment inside and outside the system.

Agricultural reconversion and biotechnologies

Economic adjustment policies have forced developing countries to engage in far-reaching agricultural and industrial reconversion in order to avoid breakdown in some sectors and promote competitiveness in others. Biotechnologies could help to meet reconversion requirements. In Venezuela, however, biotechnologies remain relatively undeveloped. The first difficulty to be overcome is that of developing commercially viable technologies. Another difficulty relates to the double standard applied by the industrialized countries to agricultural commercialization policies. On the one hand, these countries assert their allegiance to a free market, while on the other they adopt protectionist measures and promote the production of substitutes for tropical staples, as in the case of cane sugar and, to some extent, rice. This curtails the development of important sectors with high potential for biotechnologies in developing countries.

Priority areas

Biotechnologies in Venezuela should have, as a priority, the production of protein-rich animal feedstuffs. The change in consumers' dietary habits in Venezuela has generated a considerable expansion of the animal feed industry, especially in recent years. Venezuela imports almost all proteins for animal consumption, and soybeans and soybean meal are the two staples that have shown the largest increase in its food imports. Between 1979 and 1987, soybean imports showed an increase of more than 400%, and protein meal imports more than 300%, followed closely by sorghum, also an animal feed ingredient (see Table 1). The possibilities for growing soybean in Venezuela are scant, and it is hard to find other sources of conventional proteins for animal feeding.

Table 1. Evolution of main imported agricultural goods.

	1979			1987		
Product	Thousands	Tons	%	Thousands	Tons	%
Milk	50.330	50.563	100	92.643		145
Legumes	30.648	66.760	100	22.845	56.482	74
Wheat	141.614	718.806	100	137.143	1.079.166	150
Sorghum	67.011	516.809	100	69.256	543.036	163
Corn	83.432	447.888	100	0	0	0
Soybeans	14.880	42.341	100	45.257	218.585	516
Veget.oil	116.993	158.216	100	99.114	259.757	164
Sugar	76.387	324.036	100	21.280	130.905	40
Protein meal	46.091	167.066	100	150.267	724.627	433

Source: Oficina Central de Estadistica e Informatica (OECI) Anuarios de Comercio Exterior, 1979 and 1987.

However, there are biotechnological alternatives. These include experimentation with sugar-cane sub-products, research for the commercialization of various tropical legumes such as the sword bean *(Canavalia ensiformis)* and the recovery of certain residues from distilleries and brewweries. It is important also to consider the possibility of producing microbial proteins from ethanol, methanol and oil products.

Regarding sweeteners, Venezuela's National Plan for Biotechnology did consider substitutes for cane-sugar as a priority, and some research centres and at least one agro-industrial company are involved in its implementation. Although it is true that there has been a sugar deficit in the country, imports have decreased lately (see Table 1). In all probability Venezuela could become self-sufficient with respect to cane sugar, a tropical product it once even exported, and it would be inadvisable for the food industry to depend on sweeteners obtained from sources other than cane sugar. The country should therefore avoid imports of other sweeteners such as high fructose corn syrups (HFCS), obtained from maize starch, which could create a dependence on a product which cannot be produced locally. The potential of cane sugar is large, not only as a sweetener but also as raw material for other, especially biotechnological, products.

Table 2. Imports of biological products in 1986–1987.

	Tons		Thousands bolivares		Thousands $	
	1986	1987	1986	1987	1986	1987
Organic acids	3.231	3.708	33.549	79.129	4.455	5.493
Citric acid	2.073	2.815	23.889	65.215	3.180	4.487
Vitamin C	271	397	17.340	49.564	2.347	3.671
Enzymes	2.110	2.135	76.424	138.355	9.622	10.286
Alpha amylase	30	30	2.117	3.908	186	267
Cuajo	13	11	3.140	4.104	414	142
Papaïne	11	15	2.713	3.007	362	198
Gums and resins	343	489	23.827	58.102	2.542	3.941
Antibiotics	1.036	808	355.918	641.436	43.718	62.020

Source: Oficina Central de Estadistica e Informatica (OECI) Anuario de Comercio Exterior, 1986 and 1987.

The outlook is also very promising for the production in Venezuela of various biotechnology-derived products which are at present imported. These are mainly organic acids, especially citric acid, vitamin C, enzymes, gums and resins, all important for the agro-industrial sector. The data in Table 2 show that imports of organic acids, in 1987, reached a total value of $ 5 million; of these $ 4,5 pertain to citric acid. For vitamin C and its derivatives, the amount is approximately $ 3,7 million, for enzymes $ 10,3 million, for gums and resins $ 3,9 million. These amounts are significant and underline the existence of an important market.

Venezuela lacks the necessary resources to develop a biotechnological industry based on starches. Venezuela still imports about half the cereals it consumes. In the case of glucose, however, the country has a better chance due mainly to the potential of sugar-cane production which offers sub-products such as molasses and bagasse.

Besides the sugar industry, the beer and soft-drink industries are highly developed. The beer industry and distilleries yield a series of sub-products and residues – lees, bran from maltose, barley, yeast and others – which could be used to develop biotechnological industries that would produce foods at present imported, and also to reduce protein-rich animal feedstuffs. The beer and other alcoholic beverage industries, have a high level of concentration, and use conventional biotechnologies.

A similar case could be made for the soft-drink and fruit juice industries, closely related to the sugar industry, due to the high input of sugar. These industries also require high inputs of biotechnological products such as organic acids and enzymes. It is imperative to keep cane sugar as the raw material for the above-mentioned industries, and not to replace it by other sweeteners.

Conclusion

The key issue for Venezuela is not whether it should participate or not in bio-industry, but rather how and where to do so. One of the main strategies should be the development of an agro-industrial sector based on biotechnologies. In this respect, one should consider not only the present state of biotechnologies in the world, but also other aspects such as available and potential skills, the size of the market (local market and exports), investment requirements, the availability of raw materials, and the substitution of imported goods. In order to make bio-industry feasible, Venezuela would have to adopt and apply new policies concerning industrial ownership, foreign investment, technological transfer, financing and the co-operation between research centres and industries.

References

ARROYO, G.; WAISSBLUTH, M. 1988. *Desarollo biotecnolígico en la producción agroalimentaria de México: orientaciones de politica. United Nations Economic Commission for Latin American and the Caribbean (ECLAC), mimeograph.*

BUTTEL, F. 1986. Biotechnology and the Future of Agricultural Research and Development in Latin America and the Caribbean. Cali, Colombia, Centro Internacional de Agricultura Tropical (CIAT).

BYÉ, P.; MOUNIER, A. 1984. *Les futurs alimentaires et énergétiques des biotechnologies.* Paris, Institut de sciences mathématiques et économiques appliquées (ISMEA).

COMISIÓN NACIONAL DE INGENIERIA GENÊTICA Y BIOTECNOLOGÍA. 1985. *Programa Nacional.* Caracas.

COMMISSION OF THE EUROPEAN COMMUNITIES. 1988. *Propuesta de decisión del Consejo para adoptar un primer programa plurianual (1988–1993) de investigacion agroindustrial y desarrollo tecnologico basado en la biotecnología.* Brussels, 9 February.

DE JANVRY, A. *et al.* 1987. *Technological innovations in Latin American agriculture.* San José, Costa Rica, Instituto de Investigaciones de la Escuela de Ciencias (IICA).

HERNANDEZ, J.L. 1987. *El estado y la política agrícola.* Caracas, Comision Presidencial para la Reforma del Estado (COPRE).

SASSON, A. 1984. *Biotechnologies: Challenges and Promises.* Paris, United Nations Educational, Scientific and Cultural Organization (UNESCO), 315 pp.

The socio-economic impact of biotechnologies on China's rural development

Xu Zhao-Xiang

Introduction

From the late 1970s onwards, the development of biotechnologies in the advanced countries aroused much interest and concern in China, as an area of severe challenge and new opportunities in a world of rapid technological change. After years of study, policies for the development of biotechnologies in China were drafted and finally approved by the State Commission and published in the Government's Science and Technology Blue Paper no.3. Biotechnologies are now one of the top priority areas both in China's Seventh Five–Year Plan (1986–1990) and Long-term High–Technology Research and Development Programmes (Xu and Zhou, 1989).

Development of China's rural economy and the urgent demand for biotechnological innovations

China is a developing country with 22% of the world population but less than 7% of the globe's cultivated farmland. Nearly all of the arable farmland has been cultivated while the total acreage is decreasing year by year due to population growth and industrial development. For many years China put emphasis on grain production only, while avoiding other economic activities in the rural areas. As a result, labour productivity was very low and China's *per capita* annual production of grain stagnated around 300 kg during the period 1957–1977. It was not until the early 1980s that China's economic restructuring and management system reforms enabled agricultural production to develop vigorously in a very short period of time. As the labour / output ratio decreased, tens of millions of surplus farm workers were transferred from growing crops to other lines of work including forestry, fishery, animal husbandry, trade, various agricultural sidelines and rural industries.

Since 1979, newly established township enterprises have created jobs for more than 80 million surplus farm workers. While the contribution of the township enterprises to the rural economy is well recognized – their 1989 output accounted for 60% of the total product of China's rural areas – there is a lingering concern that the unplanned development and poor management of the vast number of rural enterprises may bring about further deterioration of the local natural resources and ecological environment. It is becoming increasingly imperative that there should be a proper choice of technologies catering to their development needs, e.g. the processing of farm produce, the food and feed industry, waste treatment and utilization.

It is expected, therefore, that the world trend towards biotechnology innovation and diffusion will have a major impact on the restructuring of production sectors in China, especially in the rural areas, and on the redeployment of surplus farm workers.

National priorities

The Chinese Government has listed the following biotechnology research-and-development priorities.

THE HIGH-TECHNOLOGY RESEARCH AND DEVELOPMENT PROGRAMME

Priority is given in this long-term programme to research topics that may help to develop new varieties of high-yield, high- quality, disease-resistant and pest-resistant plants and animals (particularly hybrid rice, high protein-content wheat, disease-and pest-resistant vegetables, forage grass resistant to drought and saline-alkaline conditions, fast-growing animals, and new fish species resistant to disease and cold climates); to raise nitrogen fixation in maize, soybean and vegetables; and to increase the rate of reproduction of cows. Other priority topics include: protein engineering to open up new production methods in food processing, agriculture, chemical engineering and the pharmaceutical industry; and new types of medicine and vaccines both for human beings and animals (Science and Technology White Paper No. 2, 1987).

SEVENTH FIVE-YEAR PLAN FOR SCIENTIFIC AND TECHNOLOGICAL DEVELOPMENT (1986-1990).

Under this Plan, about 45% of the planned budget for biotechnology research and development went to supporting the development of fermentation and enzyme engineering, mostly for the purpose of technical transformation of the traditional industries, while the remaining 55% supported the development of genetic and cell engineering, mainly for the purpose of developing new biotechnological products and processes. The target production sectors are agriculture, forestry, animal husbandry, fishery, pharmaceuticals, food processing, light industries and environmental protection.

THE TORCH PROGRAMME

This programme gives technical and financial support for industrial enterprises to promote the commercialization of high-technology research and development results. A number of projects in biotechnologies are being considered as priorities (Science and Technology White Paper No. 3, 1988).

THE SPARK PROGRAMME

This programme gives technical and financial support for the rural enterprises including the establishment of demonstration plants, the provision of technical know-how, equipment and training. The 1987–1990 priority areas of the Programme cover 150 different kinds of product (or production sector) under 20 different categories, most of which are biotechnology-related (Science and Technology White Paper No. 2, 1987). They include the development of high-quality, high-yield fruits; sea- and fresh-water aquaculture; poultry farming and processing; high-quality animals and leather processing; growing and processing of special-flavour vegetables; fast-growing trees and comprehensive utilization of forest products; utilization of natural essences, perfumes and pigments; utilization of animals and plants for medical purposes; comprehensive utilization of agricultural by-products; comprehensive utilization of edible oils, grains and potatoes; development of soft drinks and grape wines; development of microbial polysaccharides, single-cell protein and special forages, additives for feedstuffs; and tissue culture technology for rapid micropropagation of flowers, plants and trees.

Opportunities and challenges

So far as China is concerned, factors favourable to biotechnology development include: the existence of a cheap labour force, a huge domestic market demand, distinctive natural resources, a certain level of endogenous science and technology, the government policy to speed up implementation of a new strategy for China's coastal regions, and to give preferential treatment to foreign investment and import of foreign technologies.

The number of higher plant species alone amounts to 27,000,of which about 92% are wild plants which may be used as an important germplasm resource to develop new varieties through genetic manipulation. The newly-built National Germplasm Resource Bank, at the Chinese Academy of Agricultural Sciences, is one of the world's largest germplasm resource banks capable of storing 400,000 to 500,000 accessions. The Institute of Microbiology of the Chinese Academy of Sciences also keeps a wide collection of micro-organisms including samples of the patented varieties. Preparations are well under way to apply safety regulations and laws for the effective utilization and protection of biological resources.

It was estimated in 1988 that about 4,000–5,000 science and technology personnel were engaged in biotechnology-related research and development activities. With the recent establishment of some modern laboratories and research centres in China's key universities and research institutes, the participation of young scientists and engineers who have received advanced training at home and abroad, the carrying-out of many international co-operative activities, and the ability to tackle key science and technology problems will be further strengthened.

While long-term research-and-development programmes are being carried out at some major national laboratories, a much larger number of biotechnology-related research and extension projects are conducted in various regions, departments and enterprises. Generally the financial resources available are limited, the techniques involved are easy to handle and do not require large-scale investment, yet the combination of conventional and biotechnological techniques can benefit the rural economy, particularly in plant breeding.

Tissue culture technology has been widely used in China since the early 1970s. More than 40 varieties derived from anther culture have been developed with remarkable economic returns. The total planting area of 10 varieties of rice and wheat thus developed has been extended to more than 10 million hectares, increasing China's grain output. Further improvement has been made possible through the use of cell engineering and other techniques. The field test of a newly developed wheat variety, for example, has shown a further increase of 375 kg per hectare on a total planting area of 27,000 hectares. Protoplast culture techniques, which were thought to be difficult to apply in breeding new varieties of beans and cereal crops, have now proved successful in breeding new varieties of rice, maize, wheat, sorghum, soybean, red bean.

Rapid technological changes world-wide also imply the following challenges:

Investing in people. Given the huge population with relatively low cultural, educational, scientific and technological levels, the transition from a traditional self-contained, simple rural economy to a much more diversified and technology-based market economy calls for the creation of a vast contingent of managerial personnel as well as a new generation of farmers, workers and intellectuals.

Creation of a more favourable environment for technological innovation and diffusion. Technological innovation usually involves multi-level activities, which implies the need for integration and co-ordination of the efforts of various disciplines, institutions and government agencies. The success of the national innovation policies is due more to the correct combination of policy instruments than to the presence of any specific instrument alone.

Strengthening of endogenous capability, not only in terms of national research-and-development efforts and extension services for industry and agriculture, but also in local areas and rural enterprises. During the 1990s, a major effort should be made to build endogenous innovative capability which will help provide a closer relationship between the research and production systems, and also the endogenous capability for the management of technological changes in China's rural areas.

Major effects of the application of biotechnologies on China's rural development

A careful analysis of the state of biotechnology research and application in China might lead to the expectation that, by the year 2000, China will still be applying such techniques as haploid plant breeding and tissue culture to meet the diversified demands in rural development; it may be able to catch up with the level of developed countries in such areas as cell fusion, embryo transfer technology, hybridoma techniques and genetically engineered vaccines; in recombinant DNA manipulation of plants, the present biotechnological gap with more advanced countries would be narrowed to a certain extent; and in some cases important breakthroughs might be achieved.

Regarding the impact of these technologies on China's rural economy, the most remarkable economic benefit will come from improvements in plant breeding, especially the development and extension of the use of high-yield, highly-resistant crops such as rice, wheat, cotton and tobacco. Some newly developed, genetically engineered species should soon pass risk assessment and be put into large-scale production according to safety regulations which are expected to be promulgated in the next few years.

Under the present policy of encouraging diversified economic activities, there will be considerable expansion of large-scale commercial production of some important fruits, vegetables, economic crops and ornamental plants, through the wide application of rapid micropropagation and virus-free methods.

Embryo transfer technology has been successfully applied in animal husbandry. From 1986 to 1989, some 1,200 dairy cows received such transplanted embryos. External fertilization has also produced 'test-tube calves' and 'test-tube pigs'. It can be expected that this technology will develop, thus leading to the commercialization of embryo production including its storage, transportation and transplantation.

There is also considerable potential for the development of aquaculture in China. More than one million hectares of tidal flats suitable for aquaculture are yet to be exploited, not to mention other fresh-water and coastal water areas. Biotechnologies will play a crucial role in providing, for example, new breeds of algae of high economic value, fish or other aquatic products which are resistant to diseases or some adverse environments.

Other biotechnology-based productive sectors which may also be created or greatly developed in the 1990s include the application of hybridoma techniques to produce various kinds of monoclonal antibodies and other biological reagents for rapid diagnosis or treatment of the diseases of livestock, poultry and aquatic products; the production of genetically engineered vaccines, growth hormones, microbial pesticides, animal-use antibiotics, feed additives and plant growth regulators.

Biotechnologies will also contribute to the structural changes of the rural industrial sectors. Food-processing and fermentation industries will be greatly developed during the 1990s. In China, the output ratio between the primary products of agriculture and the processed products was only 1:1 before 1985, while in countries with modern agricultural structures, the ratio is usually in the range of 1:4 or 1:6. Biotechnologies will provide major food ingredients and various additives, as well as highly efficient biocatalysts and bio-reactors to transform the raw materials into high value-added products. Not only can the vast number of traditional industries, e.g. wine production and soy sauce manufacture, be upgraded through the use of new biotechnological achievements, but many new productive sectors will be created or further developed. These include the production of enzymes and of amino-acids, single cell proteins, food additives and other products which use various renewable resources as raw materials.

An important area of application is the treatment of industrial or farm waste to produce biogas. For a farm raising animals and poultry, wastes can be treated in anaerobic digesters to produce biogas while the solid residue of the digesters can be used again to feed poultry or fish, or as fertilizer. Waste-water from many food, sugar-making, wine-making and other fermentation processes can also be treated in anaerobic digesters to produce biogas. It is well-known that due to acute shortage of fuel in China's rural areas, biogas producers were popularized to a great extent as early as the 1970s. Up to 1979, 7 million small-scale producers were installed by the rural households. Now, with the development of rural enterprises, and the increasing amount of wastes available, much larger anaerobic digesters can be used by rural enterprises and it is much easier for the large-scale digester to upgrade its technology than the small one run by individual households.

Prospects for international co-operation

There is much room for international co-operation in the following areas:

- the 'open laboratories' established in the late 1980s with very modern facilities can help accomodate more foreign visiting scholars wishing to participate in joint basic research in China; and still more Chinese students and scholars will go abroad to receive advanced training in modern biotechnologies;
- improvement of production capability in some technology-intensive or capital-intensive sectors, by stimulating technology inflow together with direct foreign investment, i.e. setting up joint ventures with foreign firms which have the expertise in research and development, production, management and international marketing. With further developments in China's research and development programmes, there are numerous opportunities for carrying them out with the collaboration of foreign partners.

Conclusion

Judging from China's present policy in biotechnology development, it can be expected that the country would be able to benefit from biotechnologies, particularly with regard to its rural development. Of particular importance are their contribution to : meeting the basic food needs of the people; restructuring production sectors in rural areas, with the concomitant redeployment of surplus farm workers. Although the present impact of biotechnologies on China's macro-economy is still insignificant, in some districts and production sectors the impact has already been considerable, and should be much more profound in the next century.

While there is ample justification for international co-operation to speed up the process of biotechnology innovation and diffusion in developing countries, there is also fierce competition in which the developing countries are in an inferior situation. Considering the ever-increasing trend towards privatization of biotechnology development, it has become even more imperative to build up endogenous capacity to select technological options, address priority needs, absorb foreign scientific and technological achievements and adapt them to local conditions.

References

Major Topics for Biotechnology Research and Development. 1987. China's Science and Technology White Paper No. 2, Div.IV, Chapter I, Section II, State Science and Technology Commission. 153 pp.
STATE SCIENCE AND TECHNOLOGY COMMISSION. 1987. *Technology Development for Priority Production Sectors under Spark Programme.* China's Science and Technology White Paper No. 2, Div.IV, Chapter II, Section II, 163–7.
STATE SCIENCE AND TECHNOLOGY COMMISSION. 1988. *Policies for the Development of Biotechnology.* China's Science and Technology Blue Paper No. 3 (in Chinese only, an excerpt was published in China's Science and Technology White Paper No. 3, English version).
STATE SCIENCE AND TECHNOLOGY COMMISSION. 1988. *The Torch Programme.* China's Science and Technology White Paper No. 3, Div.IV, Chapter I, pp. 245–50.
XU ZHAO XIANG; ZHOU YONGCHUN. 1989. *Biotechnology in China.* Institutional reforms and technological innovation. Paper presented at the International Symposium on Biotechnology for Sustainable Development (Feb.27–March 1989), United Nations Environment Programme (UNEP), Nairobi, Kenya; *Biopolicy Series,* No. 1 of the International Federation of Institutes for Advanced Study (IFIAS).

Biotechnology research and development in the Republic of Korea

Hong-Ik Chung

Introduction

Although the Republic of Korea has a substantial food industry with a long tradition in fermentation and a fairly advanced pharmaceutical industry expanding fast since the 1960s, it was not until the early 1980s that a systematic effort at developing biotechnologies was made at national level (Moon–je Cho, 1990).

A number of scientists working in universities and government research institutes after receiving advanced training abroad in fields such as molecular biology, genetic engineering, and microbiology, were able to convince policy–makers in government and industries of the urgent need to designate biotechnologies as a national priority.

The first achievement was the establishment in 1982 of the Korea Genetic Engineering Research Association (KOGERA), a consortium of industrial firms with active interests in biotechnologies. In the same year, the Ministry of Science and Technology included biotechnologies in the list of seven strategic areas of national technology development (Moon H.Han, 1990). The Korea Research Council for Applied Genetics was then created to promote and co–ordinate research activities in universities.

An important milestone in government policy for biotechnologies was the passing of the Genetic Engineering Promotion Law in 1984, declaring the government responsible for the development and industrialization of genetic engineering, establishing a national centre for genetic engineering research and prescribing the duties of various agencies for the promotion of genetic engineering.

The national research-and-development expenditure reached US $60.7 million in 1989, nearly seven times that of 1982 (*Korean Economic Yearbook*, 1990). There are now 620 national staff with Ph.Ds and 1,798 with a master degree working in biotechnology research and development. The present level is hardly sufficient, however, as indicated by the nation's reliance on foreign biotechnology products. In biotechnology-based medical and pharmaceutical products, foreign imports accounted for 89% of the total sales valued at $ 18.9 million in 1989.

Government agencies

MINISTRY OF SCIENCE AND TECHNOLOGY

Under the general supervision of the Council for Science and Technology and the Economic Planning Board, the Ministry of Science and Technology is the most important ministry involved in biotechnology research and development. It supports biological research programmes in government research institutes, university research programmes and commercial research and development in industry. In 1987, it adopted the Long-Term Science and Technology Development Plan Toward the Year 2000 in which biotechnologies were pinpointed as a priority development area. The Ministry has 21 government research centres under its administrative guidance and provided 64% of all the government research-and-development expenditure between 1982 and 1988.

MINISTRY OF EDUCATION

The Ministry of Education is a major funding source for universities. It is primarily responsible for basic research and manpower training in biotechnologies. The Ministry awards grants annually through two foundations under its administrative responsibility, the Korea Science Foundation which provided $ 340,000 for 54 university projects in 1989, and the Korea Higher Education Foundation which allocated $ 457,000 to 55 projects in the same year (*Genetic Engineering Quarterly*, 1989).

The Office of Rural Development in the Ministry maintains ten research institutes which had a biotechnology budget amounting to US $ 1.41 million in 1989.

OTHER MINISTRIES

The Ministry of Health and Welfare provides grants for basic research in universities as well as direct support for two research institutes for biotechnology research. The Ministry of Commerce was successful in establishing in 1989 the Production Technology Research Centre.

Government research institutes

MINISTRY OF SCIENCE AND TECHNOLOGY

The Genetic Engineering Centre (GEC), affiliated to the Korea Institute of Science and Technology, was established in 1985 as the national centre for biotechnology research. Staff numbers have increased from 30 to 144, of whom 70 have Ph.Ds. The budget for 1989 was $ 6.67 million and $ 12.79 million for 1990. The Centre includes three research groups of 19 laboratories working in such fields as molecular genetics, protein chemistry, microbial ecology, microbial genetics, biocontrol, enzyme engineering and food resources systems; it also has four programmes: a gene bank; biopotency evaluation; biological reagents; a pilot plant. The GEC recently invested $ 257,000 in a joint research programme with foreign institutes: the National Institutes of Health in the USA, the Pasteur Institute in France and the Physical Chemistry Institute in Japan were its research partners in 1989. The GEC also plays an important role as a funding organization, notably providing research grants to universities and organizing seminars, conferences, and workshops.

Several other research organizations – the Korea Chemistry Research Institute, the Korea Energy Institute, the Korea Energy and Resources Institute, and the Korea Ginseng and Tobacco Institute – spent a total of $ 2.3 million on biotechnology research and development in 1989.

MINISTRY OF AGRICULTURE AND FISHERIES

Under this Ministry, the Agricultural Science Institute has a staff of 264 including 36 with Ph.Ds, and had a biotechnology research and development budget of $ 510,000 in 1989; the Plant Research Institute carried out seven biotechnology projects in 1989, with research and development spending of $ 214,000; and nine other research organizations, including the Livestock Hygiene Research Institute, were involved in biotechnology research and development in 1989, with a total outlay of $ 690,000.

MINISTRY OF HEALTH AND WELFARE

The National Health Institute conducted 10 biotechnology-related projects in 1989, ranging from the treatment of the acquired immunodeficiency syndrome (AIDS) to monoclonal antibody development, with a total research budget of $ 283,000. The Institute for Health Safety Research, established in 1987, has research programmes on testing of new medical and pharmaceutical products. It allocated $ 668,000 to 24 biotechnology-related projects in 1989.

Universities

There are 103 universities and colleges in Korea, most of which have some of the departments traditionally related to biotechnologies, e.g. biology, biochemistry, physical chemistry, food sciences, chemical engineering, agricultural chemistry. Twenty-three of these have added new departments (e.g. genetic engineering, bio-engineering, molecular biology) in biotechnologies since 1983.

Currently, 17 universities maintain biotechnology research centres, which received a total of $ 1.39 million from the government for biotechnology research in 1989. Some universities have created inter-university research centres to promote collaborative research at regional and national levels (*Science and Technology Yearbook,* 1989).

Industry

Industrial research and development in biotechnologies is led in Korea by 19 companies which have joined together to form the Korea Genetic Research Association (KOGERA). These include 7 pharmaceutical, 6 chemical, 4 food-processing and 2 textile companies. KOGERA's activities cover: development of new products and processes; collection and dissemination of information; recommendation of policies to government; training of manpower for member companies. The Ministry of Science and Technology initially provided up to 70% of the research fund which is now covered mainly by members' contributions (Organisation for Economic Co-operation and Development, 1988).

The member companies participate in the government/industry co-operative research programme as well as conducting their own research. They spent $ 29 million on biotechnology research and development in 1989, a tenfold increase from 1982. Two have also established research companies abroad. About 30 other companies outside KOGERA are also active in biotechnology research and development.

Conclusions

The Korean Government has established biotechnology development as a national priority and promoted it through policy measures ranging from funding basic research to creating venture capital organizations. Despite its success in implementing individual programmes, however, it has failed until now to elaborate a coherent national policy which, along with close inter-agency co-ordination, is of particular importance in Korea where available funds and experienced scientists are in short supply.

Biotechnology development in Korea depends on a tripartite system of research and development based on the universities, government research institutes and industrial research. In theory, the universities concentrate on basic research, government research institutes on applied research, and industry on mass-production technologies. This division of responsibilities became increasingly difficult to maintain, however, especially in the mid-1980s when the research-and-development outlays from private industry outstripped government funds.

Nevertheless, the universities have been largely successful in expanding the pool of scientists and the government research institutes, and industrial laboratories have developed a limited but important series of new products and processes.

The effectiveness of a national research-and-development system ultimately depends on three factors:

- government policy, i.e. a coherent strategy (for various reasons, Korea has had only a limited success in this area);
- the expertise and availability of qualified manpower (Korean universities have readjusted traditional curricula and created new departments in biotechnologies);
- the availability and management of funds (Korean investment in biotechnologies is small compared to more advanced countries, but has continued to

increase rapidly during the 1980s; the outlook is optimistic because industry now covers more than two-thirds of the national expenditure on biotechnology research and development).

References

Education Yearbook,1989. Seoul, Ministry of Education.
Genetic Engineering Quarterly, 1989. Korea Genetic Engineering Research Association.
Korean Economic Yearbook, 1990. Seoul. Federation of Korean Industries.
MOON, H. HAN. 1990. *Present Status of Research and Development Endeavour of Genetic Engineering and Biotechnology in Korea.* Paper presented at a Seminar on biotechnology (Seoul, March 1990).
MOON-JE CHO. 1990. Present Status of Genetic Engineering and Biotechnology in South Korea. *Critical Reviews in Biotechnology*, 10-1, pp. 47-67.
ORGANISATION FOR ECONOMIC CO-OPERATION AND DEVELOPMENT (OECD). 1988. *Biotechnology and Changing Role of Government.* Paris, OECD.
Science and Technology Yearbook,1989. Seoul. Ministry of Science and Technology.
YUAN, R.T. 1988. *Biotechnology in Singapore, South Korea, and Taiwan.* London, Stockton Press.

Contribution of biotechnologies to sustainable rural development in developing countries: a case study in Thailand

S. Bhumiratana

Introduction

Although strongly motivated to develop their resources by using biotechnologies, developing countries face a number of obstacles, the most important of which are lack of trained manpower and a concurrent or resultant weak scientific and technological infrastructure. For these developing countries, biotechnologies offer both threats and opportunities (*The UNESCO Courier,* 1987; Ahmed, 1988; Dembo and Morehouse, 1987; Yuthavong and Bhumiratana, 1988). The threats include substitution of new products for the conventional ones which have long been a source of income for the developing countries; one obvious example is the effect on the sugar industry of the new sweeteners produced by biotechnologies (Van den Doel and Junne, 1986). Opportunities are offered by the fact that many developing countries are situated in the tropical belt, often rich in natural resources which biotechnologies can convert into high value-added products. These opportunities can only be taken advantage of successfully if accompanied by increased capability in the technologies (McConnell *et al.,* 1986). Good policies and strategies for development and utilization of biotechnologies can therefore have a significant impact on the future of a developing country (Yuthavong *et al.,* 1984; Yuthavong, 1987; Bhumiratana, 1989; TDRI, 1989).

Table 1. a. Income distribution in Taiwan

	1964	1970	1975	1980	1985	1990
Richest 20%	41.1	38.7	37.9	36.8	37.6	38.2
2nd richest 20%	20.0	22.5	22.3	22.6	22.9	22.6
3rd richest 20%	16.6	17.1	17.3	17.7	17.5	17.4
4th richest 20%	12.6	13.3	13.6	13.9	13.6	13.5
Poorest 20%	7.7	8.4	8.9	8.8	8.4	8.3
Gini coefficient	0.360	0.321	0.312	0.303	0.317	0.322

Sources: Kowin Chang and DGBAS

b. Income distribution in the Republic of Korea

	1965	1970	1975	1980	1982	1984
Richest 20%	41.8	42.62	46.3	45.4	44.3	42.3
Poorest 40%	19.3	19.6	16.8	16.1	17.1	18.9
Gini coefficient	0.343	0.332	NA	0.389	0.357	NA

Sources: Economic Planning Board

c. Income distribution in Thailand

	1975/76	1980/81	1985/86
Richest 20%	49.26	51.47	55.63
2nd richest 20%	20.96	20.64	19.86
3rd richest 20%	14.00	13.38	12.09
4th richest 20%	9.79	9.10	7.87
Poorest 20%	6.05	5.41	4.55
Gini coefficient	0.426	0.453	0.500

Sources: TDRI

Importance of rural development

Economically, Thailand has enjoyed one of the fastest economic growth rates during the past half decade. Since the first Economic and Social Development Five-Year Plan which began in 1961, to the end of the plan which ends in 1991, Thailand has an average annual GNP growth of about 7%. Thailand's industrial expansion has been import-dependent, even in the export-oriented sectors, with the result that an important fraction of Thai industries has grown totally independent of the country's resources. This rapid economic development does not necessarily imply equal benefit for all, however, and a number of major obstacles are still preventing Thailand from becoming a Newly Industrialized Country (NIC). These range from inadequate local saving and poor income distribution to shortage of skilled scientists and engineers, and the lack of an indigenous scientific and technological base. In Thailand where approximately 70% of the population still inhabits rural areas, the economic growth pattern during the last two decades indicates an increase in urban-rural disparities.

It is apparent from the Gini coefficient (Bangkok Bank Report, 1989) that Thailand ranks far behind other newly industrialized countries (see Table 1), and that the gap between the richest and the poorest is widening, contrary to trends in other NICs.

Much attention was to providing a model for future development. Given the country's potential, well balanced growth in three sectors – industry, agriculture and services – is required. The recent industrialization did not place sufficient emphasis on the interdependence of these three sectors, with a resulting increase in capital-and-labour intensive industries as opposed to a knowledge-intensive industry. The three sectors linked together should change their respective capacities to employ manpower in a balanced fashion.

Status of biotechnologies

Three major reasons for Thailand's active interest in biotechnologies are: it is a fertile country with a vast pool of unused natural resources; for a developing country of its size, Thailand has a relatively large number of qualified research personnel for biotechnological development; and it has a fairly sound economic status, which would ensure returns in judiciously chosen biotechnology investments. The last two decades in particular have seen a remarkable growth in basic life sciences in Thailand, linked with the emergence of a critical mass of scientists trained in developed countries and the establishment of research–based graduate schools providing substantial indigenous research capability. During that same period, problems in agriculture and rapidly expanding agro-industry, health and the environment have made it necessary to make use of biotechnologies. Adequate linkages are, however, still lacking between the research and university sector, and the private and the government sector (Yuthavong *et al.,* 1984; Yuthavong, 1987; Yuthavong and Bhumiratana, 1988; Bhumiratana, 1989; TDRI, 1989).

One factor that helped to strengthen biotechnologies in Thailand was the establishment of the National Center for Genetic Engineering and Biotechnology (NCGEB) in 1983. The NCGEB's primary objectives are to act as the focal point for strengthening Thailand's capabilities in genetic engineering and biotechnologies and related areas, and to apply these to national economic and social development. The Center formulates policy and plans on biotechnologies, provides support for research–and–development activities in designated institutions and links these institutions with the private sector. The establishment of the National Center marked, for the first time, significant local funding from the government towards development of a specific technology. Support for the technology is, moreover provided as a complete package with the provision of funding, information, training, links to industry as well as international linkages. The National Center has 5 affiliated laboratories, including pilot plants, and over 66 projects in 10 institutions in the network (Yuthavong and Bhumiratana, 1988; NCGEB, 1990). The emphasis is on the development, transfer and utilization of biotechnologies, including genetic engineering, in industry, agriculture, public health, energy and environment, and on the strengthening of basic infrastructures. The Center also commissions studies on the status of specific technologies or industries, to assess their economic and social importance and to pinpoint the research, development and technology transfer needs to be met.

The interests of Thai industries in relation to biotechnologies include the areas of amino–acid production for feedstuffs, cassava starch modification, hybrid seed production, commercial plant propagation through tissue culture, secondary production of antibiotics, animal vaccines, beverages and foodstuffs. The private sector mostly acquires technologies through import, and it is hoped that the future supply of local biotechnology expertise will help the choice, acquisition and development of the imported technologies. A recent study identified several major constraints to the technological development of the industry (Thailand Development Research Institute, 1989), which included the inadequacy of science and technology manpower, the lack of private sector interest in research and development, the inacessability of appropriate technical information, weak technical consultancy and inadequate public–private sector linkages.

Another recent boost for biotechnologies, as well as for other frontier technologies in Thailand, is the launching of the Science and Technology Development Project, through the USA–Thailand bilateral government co-operation programme. The activities to be implemented include the provision of research-and-development grants in biotechnologies, and of grants and soft loans for bio-industrial development, and the promotion of linkages between the universities and the private sector. In order to ensure successful development of bio-industries, the Thai Government is in the process of reforming investment and taxation regulations.

Rural development and biotechnologies

The Thailand Rural Development Projects were started in the early 1980s in order to alleviate rural hardship, to increase agricultural production through diversification, and to provide pilot studies on agro-processing industries in rural areas and their socio-economic problems. Processing plants were set up as a core for the project in remote regions in northern and north-eastern Thailand. A scheme for guaranteed crop prices was set up to ensure at least a minimum level of return for farmers while still maintaining free market forces. Emphasis is placed on the farming or agricultural extension aspects rather than the processing, in contrast to most agro-industrial enterprises.

Pilot rural development projects starting with land and water development and management, followed by crop production and culminating in agro-industry are to be established in various parts of the country. Currently the projects support supplementary income activities for approximately 15,000 small farming families who would otherwise find it difficult to develop beyond subsistence. Each family (on average 6 persons) owns its small farm of approximately 2–5 acres which can be under activity at any given time. For the year 1990, cash returns to these small farming units total more than $ 2 million. Experience gained has provided the following insights.

1. It is essential that any rural development project must be sustainable. For this reason, biotechnology applications must be cost-effective; this is not always easy to achieve, as they are usually associated with research-and-development activities which tend to be expensive, and as the techniques

required are normally available only in large and often private laboratories. If biotechnologies are to be effective in current practice, appropriate means for the transfer of technology for specific produce such as "baby corn", straw mushroom and tomatoes, must be sought.

2. To maintain a viable rural population the effectiveness of small farmers as producer units, as compared to fully or partially mechanized large farms where farmers become labourers, must be improved. This requires a different development approach. Funding is another major constraint both to utilizing costly research results and to carrying out further development work.

3. In attempting to minimize the cost of production while maximizing farm yields, the use of agricultural chemicals is widely practiced and in many cases this is not done properly. Health and environmental considerations are for all practical purposes neglected. Biotechnologies could solve this problem.

While biotechnologies hold much promise as appropriate technologies to accelerate the development process, experience from Thailand shows that participation by all sectors, from research to end-users, is required before the potential of these as yet "expensive" technologies can be fully realized.

References

AHMED, I. 1988. The Bio-Revolution in Agriculture: Key to Poverty Alleviation in the Third World? *International Labour Review,* 127, pp. 53–72.

BANGKOK BANK REPORT, 1989. *What it Takes to Be a Newly Industrialized Country.* Research Office of the Bangkok Bank. March 1989.

BHUMIRATANA, S. 1989. Biotechnology Education and Training in Developing Countries. In: *Regional Seminar on Public Policy Implications of Biotechnology for Asian Agriculture (New Delhi, India, 6–8 March 1989).*

BHUMIRATANA, S. *1990. Thailand's Science and Technology Capability. In: Workshop on Science and Technology Development in ASEAN Countries: Problems, Prospects, Scope for Co-operation with European Countries.* (Bangkok, Thailand, 7–8 June 1990).

DEMBO, D.; MOREHOUSE, W. 1987. *Trends in Biotechnology Development and Transfer.* Vienna, United National Industrial Development Organization (UNIDO), 96 pp.

MCCONNELL, D.; RIAZUDDIN, S.; ZILINSKAS, R.A. 1986. *Capability Building in Biotechnology and Genetic Engineering in Developing Countries.* Vienna, United Nations Industrial Development Organization (UNIDO), 114 pp.

THAILAND DEVELOPMENT RESEARCH INSTITUTE (TDRI). 1989. *The Development of Thailand's Technological Capability in Industry. Vol. 6: Overview and Recommendations.* Bangkok, TDRI.

UNESCO, 1987. *The Courier,* March 1987. Paris, United Nations Educational, Scientific and Cultural Organization.

VAN DEN DOEL, K.; JUNNE, G. 1986. Product Substitution through Biotechnology: Impact on the Third World. *Trends Biotechnol.,* No. 11, pp. 88–90.

YUTHAVONG, Y. 1987. The Status and Future of Biotechnology in Developing Countries.

In: Vasil, I.K. (ed.). *Biotechnology Perspectives, Policies and Issues,* pp. 127–43. Gainsville, University of Florida Press.

YUTHAVONG, Y.; BHUMIRATANA, S. 1988. National Programs in Biotechnology for Thailand and Other South-East Asian Countries: Needs and Opportunities. In: *Proceedings of the Conference on Strengthening Collaboration in Biotechnology: International Agricultural Research and the Private Sector.* (USAID, April 1988).

YUTHAVONG, Y.; BHUMIRATANA, S.; SUWANNA-ADTH, M. 1984. The Status and Future of Biotechnology in Thailand. In: *Proceedings of ASEAN-EEC Seminar on Biotechnology: The Challenges ahead.* (Singapore, November, 1983), pp. 29–30. Science Council of Singapore.

Conclusions

A. Sasson

General considerations

Biotechnologies are taking root in developing countries, in some of which they have already reached a rather advanced stage of development, it being understood that developing countries correspond to different economic, social and technological situations; biotechnologies offer a wide range of tools, with various levels of technical sophistication, economic investment and efforts; hence the possibility of choosing the appropriate techniques, with a view to achieving the most positive impacts. Generalization should be avoided and solutions adopted which meet the specific needs of each country or region.

The various geopolitical or geocultural regions show striking differences in the adoption, adaptation and application of biotechnologies. The countries of the Pacific Rim and South–East Asia, including the so–called Newly Industrialized Countries (NICs), have given biotechnologies a high priority in their development strategies, as they anticipate a positive economic, and perhaps social, impact. Sub–Saharan Africa, on the other hand, except a few "pockets", is far behind, which is a major source of concern for the international community.

By far the most widely used biotechnologies in developing countries are those relating to plant tissue culture, *in–vitro* micropropagation and massive clonal multiplication of food and horticultural crop species, flowers, forest and fruit trees, and plantation or estate crop species. Hence, the impact on agriculture, food production and exports of agricultural products, as well as on a large sector of the economy of developing countries. Food fermentations and processing, livestock husbandry, bio–energy production, aquaculture, and animal and human health receive increasing attention, and their social and economic impacts are important.

Whatever the present or anticipated impacts, biotechnologies are neither a panacea, nor a substitute, to conventional and effective techniques such as those of crop breeding, animal breeding, or traditional vaccine production. They are new inputs, new tools, and therefore improve and enhance the impact of current technologies. Biotechnologies cannot develop in a vacuum: for example, to be effective plant biotechnologies need to be supported by agricultural research and extension.

IMPACT STUDIES

It seems that social scientists have not yet extensively studied the area, although case studies on specific impacts exist, such as those of the International Labour Organization (ILO) on rural labour and employment, or those regarding the effects of sucrose substitution by high fructose corn syrups. There are, however, numerous publications on the possible impacts, including warnings about negative impacts in developing countries, mainly regarding the role of multinational corporations and the private sector, the market-driven features of many biotechnologies, the experience of industrialized countries, as well as to the impact of the 'green revolution' on the farming sector of developing countries.

More focused studies are certainly needed at national level. This includes the collection of segregated data, in order to distinguish the actual impact of biotechnologies from that of other technologies applied in the same sector, e.g. agriculture, health, animal husbandry, bioremediation, and to establish which techniques have proved most successful and whether they may be adapted to the region in question.

POSSIBLE ECONOMIC IMPACTS

It is possible, however, to deduce economic impacts, in the positive sense, from the following applications in several developing countries:

- creation of new rice varieties through anther culture and development of haploid lines, followed by diploidization and hybridization, already planted on large areas and showing a marked increase in yields (e.g. in China – an experiment which may be extended to other Asian countries);
- potato multiplication, using micropropagation techniques to produce disease and pathogen-free planting material, thus improving yields, extend-

ing the consumption of potato in the diet, avoiding imports, improving germplasm exchange (e.g. in the Andean countries, Thailand, Vietnam, Morocco and sub-Saharan Africa);
- enhancement of nitrogen fixation and increase of yields of legumes through the production of effective inoculants of *Rhizobium,* and eventually of actinomycetes and mycorrhizae (e.g. in Kenya, Senegal, Thailand, Zambia, Zimbabwe);
- improvement of horticultural crops and fruit trees; planting or replanting with pathogen-free material: banana (e.g. in Egypt and Morocco), citrus (e.g. in Morocco, the Mediterranean and Asian countries), sugar-cane (e.g. in Mauritius and the French Antilles), strawberry (e.g. in Egypt, Thailand and Zimbabwe), asparagus, artichoke (e.g. in Egypt); this provides farmers with a new source of income and consumers (local or foreign) with a wider range of products;
- expansion of the flower industry; carnations in Colombia, roses in Morocco, orchids and temperate flowers in South-East Asia (there is great demand and considerable opportunity, due to the expanding markets in the European Economic Community and Japan);
- large reforestation schemes, using plantlets or seedlings derived from tissue or organ culture (e.g. teak and several other tropical tree species, acacias, casuarinas to fix sandy soils in Morocco, Senegal and Tunisia); not only economic but also environmental impacts may be expected here ;
- estate and plantation crops, e.g. oil-palm (Ivory Coast, South-East Asia); date-palm, especially *Fusarium* wilt (bayoud) tolerant varieties or lines; tea, coffee, cocoa (Latin American countries);
- microalgae production for feed, health foods and biochemicals, e.g. beta-carotene (Thailand, Vietnam);
- food fermentations (e.g. in Indonesia tempeh production, which does not meet the demand and can be improved and extended with little investment in research and development and in industrialization; gari in Sub-Saharan Africa);
- ethanol production of which the economic, social and even environmental impacts have been well studied in Brazil.

In evaluating impacts it should be borne in mind that the application of biotechnologies is in a development phase, even though evolution may be rapid, and that failures must be accepted and carefully examined if progress is to be made.

CONDITIONS FOR A POSITIVE IMPACT

It is not sufficient to have a strong biological research base; there is also a need for a flexible approach to harness and focus capital in a productive manner, for mechanisms to transform research and development into profitable industrial ventures, and to improve options for agricultural products and raw materials.

The application of biotechnologies in developing countries cannot be organized on a national scale without strong government backing. Careful policies are required, establishing priorities and covering all factors involved, including the country's level of development, political structure, climate, agricultural traditions, food and commercial crops, etc. If policy-makers are to make judicious choices as regards biotechnologies, the dissemination of available information is essential. Relevant information concerning research and its applications, which at present tends to circulate mostly among the industrialized countries, should be channelled more effectively to the developing countries. Increased networking and exchange among existing research groups could also improve the flow of information.

The private sector should also be involved especially in research. New prospects are opening up in South-East Asia and Latin America, thanks to the role of the regional and development banks, and new ways are being found of sharing responsibilities between the public and private sectors.

Comparative advantages should be strengthened or at least the accelerated loss of these advantages lessened, by emphasizing quality standards and drawing benefit from early cropping and from environmental assets.

Increased productivity through the use of biotechnologies does not guarantee an equal distribution of income along all sections of the population. Appropriate social policies are therefore required to protect the poorer social groups and the small farmers. If biotechnologies are to be applied effectively in developing countries, the needs of the users must be taken into account, and in the case of agricultural applications, this means the small farmers. Biotechnologies can be used to help achieve lasting, indigenous and environmentally sound rural development.

Further research is needed on the impact of biotechnologies on labour absorption and employment, which could, for instance, increase with the practice of multiple cropping or the extension of cultivated land, yet could be adversely affected by other factors, e.g. the decline of the sugar export market resulting from the increased use of sugar substitutes. Although substitution does represent a real threat to some developing countries, there are as yet not many examples, apart from sugar. This is an area, however, in which vigilance is required, and research could help to ascertain which are the most threatened exports and what are the alternative uses for displaced crops or land.